Introduction to Built Asset Management

# Introduction to Built Asset Management

*Dr Anthony Higham, Dr Jason Challender and Dr Greg Watts*

WILEY Blackwell

*Registered Offices*
John Wiley & Sons, Inc., 111 River Street, Hoboken, NJ 07030, USA
John Wiley & Sons Ltd, The Atrium, Southern Gate, Chichester, West Sussex, PO19 8SQ, UK

*Editorial Office*
9600 Garsington Road, Oxford, OX4 2DQ, UK

For details of our global editorial offices, customer services, and more information about Wiley products visit us at www.wiley.com.

Wiley also publishes its books in a variety of electronic formats and by print-on-demand. Some content that appears in standard print versions of this book may not be available in other formats.

*Library of Congress Cataloging-in-Publication data*
Names: Higham, Anthony, author. | Challender, Jason, author. | Watts, Greg
  (Gregory N.), author.
Title: Introduction to built asset management / Dr. Anthony Higham, Dr. Jason
  Challender, and Dr. Gregory Watts.
Description: Hoboken, NJ : John Wiley & Sons, [2022] | Includes
  bibliographical references and index.
Identifiers: LCCN 2021033080 (print) | LCCN 2021033081 (ebook) | ISBN 9781119106586
  (paperback) | ISBN 9781119106579 (pdf) | ISBN 9781119106562 (epub)
Subjects: LCSH: Real estate management. | Commercial real estate. | Real property.
Classification: LCC HD1394 .H54 2022  (print) | LCC HD1394  (ebook) |
  DDC 333.5068--dc23
LC record available at https://lccn.loc.gov/2021033080
LC ebook record available at https://lccn.loc.gov/2021033081

Cover Design by Wiley
Cover Image: © coldsnowstorm/Getty Images

C9781119106586_040122

# Table of Contents

# List of Figures

# List of Tables

# Foreword

For too long and for too often the capital expenditure associated with construction project development has drawn attention away from the operational expenditure associated with building maintenance. Awareness of the importance of the maintenance of existing assets, and the design for maintenance of new, has risen amongst public policy professionals and owners, as well as construction practitioners. The climate emergency, targets for zero carbon and the opportunities for change driven by digital technologies are driving change in clients' needs and in suppliers' business models in the sector.

The UK Industrial Strategy – beginning with the report *Construction* 2025 – brought policy to bear on the maintenance and upgrading of existing buildings. The operation phase of a building's life cycle should no longer viewed in isolation, but instead as part of the design and construction phases. The utilisation of the knowledge and experience from building maintenance professionals during the design and construction of a building can have untold benefits during the building's operation.

This textbook provides a solid introduction to the importance of the building maintenance sector. It helps lay a foundation of knowledge for how professionals can help successfully manage the challenging and often complex nature of building maintenance. It brings effective building maintenance under the lenses of contemporary practices such as procurement and contracting, financial management, sustainability, maintenance systems, and risk management. Surveying the existing building stock is also discussed, as this is of key importance if buildings are to meet future population demands and sustainability targets.

With their diverse knowledge and experiences from a range of roles across the construction industry, the authors are well placed to address the challenges and requirements of building maintenance. This textbook provides a comprehensive introduction to, and pragmatic application of, the principles of effective building maintenance for both students and practitioners alike. For those entering the world of building maintenance it introduces and discusses both the longstanding and contemporary issues professionals face and considers how such challenges can be overcome for the benefit of future generations.

**Professor Peter McDermott**
**University of Salford**

# Acknowledgement

The authors would like to thank Ben Lyere, for his assistance with Chapter 4: Maintenance Management and Performance Measurement as Part of Private Financing Initiative (PFI) Schemes. Ben previously carried out extensive research into the challenges of performance measurement of facilities and maintenance management in a PFI hospital context and was instrumental to collating the themes and focus for this chapter of the book.

# 1

# Introduction

## 1.1 Introduction to the Book

Undertaking responsible and practical repairs and maintenance policies and strategies are becoming an ever-important factor in life cycle management of property assets from owner occupiers to large property portfolios. In this sense, building maintenance practices are needed in every development and is therein a very crucial part in every life of every building. This may be predicated on buildings requiring to be well maintained in order to retain the value of the property itself. Furthermore, the buildings will also continue to fulfil their functions if properly maintained and will give the convenience to the tenant and occupants in the buildings.

With the technological advancement in buildings over many years and the advent of more complex building elements and services, the area of building management is becoming a 'hot topic' and one that is worthy of a more profound and professional approach and address. There has become a need for more prescriptive and perhaps regimented maintenance programmes and procedures to ensure statutory compliance is maintained as legislation becomes more convoluted, and to ensure buildings do not develop defects and deteriorate.

Some earlier publications have successfully considered aspects of maintenance, including its dimensions, planning, organisation, procurement and post-contract management and finally the role of knowledge management within maintenance organisations. Despite the comprehensive coverage of the fundamental issues, most of these previous works have not considered the increasingly important areas of carbon resilience and information management, which the present book will address. This is an important and topical aspect, as it is now essential that clients need to consider the viability of retrofitting microgeneration and renewable technologies to existing structures during major maintenance activities. This book will seek to readdress the balance and will provide readers with a depth of knowledge around such areas in an attempt to improve best practice in maintenance management.

## 1.2 The Main Areas and Themes Covered in the Book

As an introduction to the book and to provide context for readers, an overview is provided below of all the main areas and themes covered alongside related discussion points.

*Introduction to Built Asset Management*, First Edition. Dr Anthony Higham, Dr Jason Challender, and Dr Greg Watts.
© 2022 John Wiley & Sons Ltd. Published 2022 by John Wiley & Sons Ltd.

### 1.2.1 Chapter 2 Surveying Existing Buildings

This chapter will outline the background to conducting building surveys and identify the skills, steps and requirements for them to be completed successfully. It will describe the methodology by which defects are correctly diagnosed through collection of evidence and observing telltale signs. Notwithstanding this premise, it will provide instances when this can sometimes be challenging, and misdiagnosis may remain a real factor. To assist readers, it will describe circumstances and practical examples where there may be more than one force of nature at play and how these can lead to different forms of building deterioration. Furthermore, it will provide examples where certain defects have been misdiagnosed for others and where symptoms are similar, e.g., alternative forms of structural cracking and dampness.

The chapter will also highlight the importance of client engagement both before and after building surveys and will describe the type and form of background information that may be useful to them. In this regard the usefulness of desktop studies and reviewing previous technical reports will be discussed. Following on from this, the importance of understanding the clients' requirements in the type and nature of information to be captured in the survey will be articulated and explained.

The different types of building survey, ranging from full structural surveys to valuation surveys, will be outlined in the chapter together with the respective benefits and possible limitations of each. In addition, the range of different surveying equipment and tools will be identified. The use and application of each piece of equipment to assist surveyors in their inspections will be covered, especially where access and visibility difficulties are experienced. Photographs of different building surveying equipment and apparatus will be provided.

The modus operandi by which surveyors will conduct their inspections will be discussed, alongside factors such as health and safety and limitations which will need to be considered. In addition, the challenges and obstacles which sometimes confront building surveyors in the course of their inspection will be identified, especially where disruption to normal building operations remains an important factor. Examples of such factors will be given and an understanding of the difficulties that these constraints can impose on the building surveying process.

Following on from this, the importance of further investigation work will be outlined in circumstances that require opening up works or more specialist input. Finally, the various stages of managing the remedial work process will be described, alongside the importance of client consultation of the different options and preferred solutions. Examples of different scenarios leading up to the agreement of the agreed remedial works will be provided.

### 1.2.2 Chapter 3 Common Maintenance Issues and Managing Defects

This chapter of the book is focused on the pathology of buildings with particular emphasis on building degradation and will describe the ways and means of identifying and addressing potential defective elements and defects within buildings. In this pursuit, it will examine the different methodologies for diagnosis of defects and their root cause (rather than just the symptoms of defects) to assist in identifying the most appropriate type of repair.

An analysis of the underlying conditions and mechanisms that lead to defects and building failures resulting from poor design and quality of build during the construction stage or refurbishment will be covered, from dampness to structural movement, corrosion and timber decay. Furthermore, the role of maintenance across the life cycles of buildings will then be discussed and the influence that this has on their preservation and deterioration. The debate around this important area will then extend into the examination of planned and reactive policies for maintenance management of buildings estates.

The challenges associated with identifying the underlying causes of building failures will be presented where one or more possible different diagnoses may be initially debated. In this context the requirements for further investigation will be examined in correctly determining the underlying root cause of problems more accurately and confidently. The different categories of building failures which lead to the emergence of defects will then be discussed and the mitigation measures that should be undertaken to reduce the risk of such failures. In addition, the individual forces at work leading to different types of defects and how these can vary depending on climate and the geographical locations around the world will be articulated. Furthermore, the difficulties and challenges of correctly identifying faults and defects to building services such as heating, lighting, fire alarms, cooling and other forms of gas and electrical installation will be explained, and practical examples provided.

The importance of understanding the nature and effect of agents that can lead to building defects will be examined in detail, including human misuse of buildings and deliberate vandalism, coupled with electromagnetic, mechanical, chemical, building user, thermal and biological related forces and effects at work. The wrong use of materials and components and design defects will be discussed and how these can develop into latent construction defects in the future, therein introducing maintenance liabilities for building owners. Thereafter, the magnitude of managing remedial works and the associated disruption that this can bring for building users will then be investigated with reference to different examples in practice. Dilemmas associated with repair or renewal decisions will be introduced and debated in detail and the overall effects that decision making can have on future maintenance and lifecycle costings analysis. This will entail a discussion around capital cost (CAPEX) versus operational cost (OPEX) perspectives. Following on from this, the various influences of different factors on the repair or renewal decision making will be covered, including capital funding, building longevity and potential disruption from proposed remedial works.

Managing the remedial works processes to address maintenance issues and defects will be examined, with a focus on best practice and achieving the correct level of supervision and quality control throughout the contract period. This will introduce quality control and best practices in selecting the right design, a team and contractors from the perspective of attaining the necessary skills, experience and competency levels required to undertake projects successfully. Finally, measures to mitigate and prevent defects will be covered and the importance of allowing access provisions in building to enable and carry out regular maintenance inspections as part of planned preventative maintenance programmes.

### 1.2.3  Chapter 4 Maintenance Management and Performance Measurement as Part of Private Financing Initiative (PFI) Schemes

The chapter will start with a background to the discussion on performance measurement (PM) of maintenance and other facilities management services in the healthcare sector. In this sense it will articulate some of the problems that PM initiatives have experienced in the past. Furthermore, it will also provide justification for highlighting the issue of performance measurement linked to PFI initiatives and explain the reasons why it is widely regarded as a 'hot topic'. Thereafter it will explore the definitions and concepts of facilities management, the application of performance measurement in a healthcare PFI setting and the impact of performance measurement on service and quality improvements. The significance, importance and challenges for PM around facilities management on a PFI scheme will then be discussed. This will include such issues as financial dilemmas facing the healthcare sector in the UK and how it has been tasked to achieve more with less, and targets to improve the quality of the health service whilst reducing cost. Other challenges for maintenance and facilities management will be covered, including the need to increase accountability, drive up quality, meet increasing demands on maintenance and facilities outcomes, make efficient savings and re-energise value-for-money approaches. In the context of such challenges further explanation will be presented on why these factors have all led to renewed interest in the direction of performance measurement. In addition, it will also explore payment mechanisms as one of the underlying reasons for measuring performance in a PFI healthcare project as a gain share/pain share initiative. In this sense, it will explain the issues around deduction of payments for poor performance and conversely financial incentives for exceeding targets. Various performance measurements for improving maintenance and facility management outcomes tools are then explored alongside the role of key performance indicators (KPIs). The benefits of performance measurement will be articulated from a position of presenting a clear picture of where improvements are actually happening.

Justification for introducing measure performance measures on PFI schemes and the need to assess progress against predetermined objectives will be covered. This will include measures to identify areas of strength and weakness and the need to align future initiatives with the aim of helping to improve the organisational performance. The chapter will then extend into performance measurement as applied to quality improvements in maintenance and facilities management practices. Financial and non-financial measurements, being the primary variables that can be used to research the construct of performance measurement in most organisations, will then be examined. The various forms of performance monitoring tools will be identified and analysed and how these relate to benchmarking and key performance indictors. Finally, key issues arising for performance management as part of a facilities management (FM) tool on PFI schemes will be articulated and discussed, alongside reflections for the future.

### 1.2.4 Chapter 5 Procurement and Contracting for Maintenance and Refurbishment Works

This chapter will look at the processes associated with the procurement and contracting of refurbishment and maintenance works. It will emphasise the importance of the way we buy from the construction industry and how this will ultimately decide the success or failure of projects. Furthermore, it will introduce strategic procurement as a tool to elicit project performance requirements and will make the case for businesslike approaches to procurement strategies. The chapter will then introduce the various procurement routes which can be adopted for refurbishment and maintenance projects before providing an overview of procurement governance requirements as projects are commissioned within the marketplace. The second section of the chapter will deal with contracts and contract selection. Given the dominance of the Joint Contract Tribunal (JCT) and New Engineering Contract (NEC) in the UK construction market, this chapter will conclude by introducing the JCT and NEC contract suites before exploring the various contract options with each suite suitable for the management of building maintenance and refurbishment projects.

### 1.2.5 Chapter 6 Financial Management: Capital Costs

This chapter will focus on the capital costs and financing arrangements of building work, including refurbishment and maintenance, which is a growing area with a predicted market value of total construction work in the UK alone of £66bn. The scope of this chapter will look at how the capital costs of such works can be managed and controlled from their inception until the point that the construction phase is completed. Furthermore, it will explain and discuss the trigger for these works from different perspectives, including from building surveys or alternatively from strategic reviews carried out on organisations' asset requirements.

This chapter of the textbook will show how pre-contract financial management develops through the option selection and design phases of projects. It will also articulate the phases of pre-contract financial management implemented by the cost consultant/quantity surveyor from the very outset of projects. As the design develops, the chapter will explain how the cost consultant/quantity surveyor will move forwards with their financial management of the pre-contract stages of the project, and how cost plans will be produced. Thereafter it will explain how cost plans provide statements of how the available budget will be allocated to the various elements of the building.

### 1.2.6 Chapter 7 Financial Management: Life Cycle Costing

This chapter will look at the growing importance of life cycle cost analysis with a major focus on predicting the costs of maintenance, occupancy and replacement of elements, sub-elements or components over many years. It will consider the longer-term interests that both clients and tier one contractors are now taking, and the paradigm shift in focus away from lowest price award mentality in favour of one led by a longer-term focus on value. It will stress the trade-off between the initial capital costs and longer-term operational costs of assets or components, especially in the public sector. As this balance continues to play out in the marketplace, the chapter will stress the importance of how future maintenance costs can be forecast in an ever-changing facilities management environment. However, it will discuss how further effort is required if the deeply ingrained business culture that compartmentalises capital and maintenance funding is to be overcome on traditional projects. In addition, it will introduce TOTEX, CAPEX and OPEX costs related to buildings and discuss the way many organisations manage their asset budgets. Finally, the chapter will introduce the New Rules of Measurement (NRM) and the life cycle cost plan.

### 1.2.7 Chapter 8 Sustainable Maintenance Management

This chapter will introduce the idea and concept of Sustainable Maintenance Management and will argue for a change in construction industry practices that embraces the role and importance of the building maintenance professional during the construction stage of works. It will explain the knowledge and experience during the capital expenditure stages that will ultimately lead to a more sustainable and efficient operational stage of a building's life. Thereafter the circular economy will be introduced and how it is challenging and changing the historically linear thinking of the built environment when it comes to the sourcing of materials.

Concepts such as carbon neutral and retrofitting will be outlined in the chapter and their importance to the role of the building maintenance professional discussed. Reporting frameworks, such as BREEAM, will also be highlighted and how they can be used for the maintenance of buildings. The corporate social responsibility (CSR) of building maintenance companies will then be discussed as well as the principles of the concept generally and how it can guide the sustainable and socially responsible behaviours of organisations. Finally, the Sustainable Development Goals (SDG) will be introduced as a method of bringing together the actions of nations, organisations and individuals to help navigate the different sustainable options that exist and attempt to focus behaviours for maximum positive impact.

### 1.2.8 Chapter 9 Risk Management

This chapter will describe and explore how building maintenance professionals encounter risk, both contractually and from the physical risks involved in the maintenance of buildings. It will introduce the concept of risk and explain the traditional view of risk which has always been somewhat pessimistic. Conversely, it will present the upside of risk and will explain how risk can provide potential opportunities rather than just threats. It will define and analyse the notion of risk and provide an exploration of how the correct actions be taken with regards to the retaining or transferring of risks. It will explain how consideration of risk in the round is essential in the current climate of the built environment when it comes to the maintenance of buildings and assets. Finally, the chapter will then discuss the tools and techniques that can be adopted by building maintenance professionals to help protect themselves (and the buildings they maintain) against excessive and unnecessary risk exposure.

### 1.2.9   Chapter 10 Managing the Maintenance Process

This chapter is focused on programming and planning, and reactive and proactive, maintenance. It will touch upon dealing with site constraints, and health and safety management when maintaining buildings and will include some elements on health and safety files and building information modelling (BIM).

It is important to have an appreciation of the maintenance process, and for the key management requirements to be understood. It is only through such an appreciation can any maintenance project be managed effectively and the intended targets achieved. The effective management of maintenance works can also lead to reduced health and safety accidents and increased building performance. Therefore, this chapter will give an overview of how to manage the maintenance process. Key areas of importance will be covered, including how to plan for maintenance works and the distinctions of proactive (scheduled) and reactive (corrective) maintenance. The importance of maintenance schedules and how they can form the basis of wider programmes of work will also be discussed, alongside the key benefits of programming and planning. In addition, the key attributes required from those collaborating in the creation and development of a building maintenance programme will be analysed.

The importance of inspecting works at the earliest opportunity and the key considerations to be made regarding site and task constraints will also be included. An overview of prevalent and successful health and safety initiatives will be provided and an introduction to, and discussion of, the benefits of adopting a soft-landing approach. Finally, the current and potentially future practices of the building maintenance professional's role will be discussed, and how by engaging them at an earlier stage in the construction process this can lead to more success in operating a building.

### 1.2.10   Chapter 11 Conclusion

The book will finish with an overall summary of the main points raised in each of the above chapters. It will provide conclusions to the themes and subject areas discussed and provide much needed reflection and recommendations in some cases for improvement in future practice.

## 1.3   Research Sources

The research for the book has been derived from a combination of sources which include the authors' own experiences, interviews with a wide range of construction professionals, and literature. The book is mainly intended for construction practitioners, including employers, design teams, contractors, subcontractors and lower levels of the supply chain. It could also be useful for teaching and learning and suit a wide target audience including under- and postgraduate students and academics. The authors are hopeful that it will make a constructive and useful contribution to the field.

# 2

# Surveying Existing Buildings

## 2.1 Introduction

This chapter will outline the background to conducting building surveys, identifying the skills, steps and requirements for successful completion. It will describe the methodology by which defects are correctly diagnosed through collection of evidence and observing telltale signs. Notwithstanding this premise, it will provide instances when this can sometimes be challenging, and misdiagnosis may remain a real factor. To assist readers, it will describe circumstances and practical examples where there may be more than one force of nature at play and how these can lead to different forms of building deterioration. Furthermore, it will provide examples where certain defects have been misdiagnosed for others and where symptoms are similar, e.g., alternative forms of structural cracking and dampness.

The chapter will also highlight the importance of client engagement both before and after building surveys and will describe the type and form of background information that may be useful to them. In this regard the usefulness of desktop studies and reviewing previous technical reports will be discussed. Following on from this, the importance of understanding the clients' requirements in the type and nature of information to be captured in the survey will be articulated and explained.

The different types of building survey ranging from full structural surveys to valuation surveys will be outlined in the chapter together with the respective benefits and possible limitations of each. In addition, the range of different surveying equipment and tools will be identified. The use and application of each piece of equipment to assist surveyors in their inspections will be covered, especially where there are access and visibility difficulties experienced. Photographs of different building surveying equipment and apparatus will be provided.

The modus operandi by which surveyors will conduct their inspections will be discussed, alongside such factors as health and safety and limitations which will need to be considered. In addition, the challenges and obstacles which sometimes confront building surveyors in the course of their inspection will be identified, especially where disruption to normal building operations remains an important factor. Examples of such factors will be given and an understanding of the difficulties that these constraints can impose on the building surveying process.

*Introduction to Built Asset Management*, First Edition. Dr Anthony Higham, Dr Jason Challender, and Dr Greg Watts.
© 2022 John Wiley & Sons Ltd. Published 2022 by John Wiley & Sons Ltd.

Following on from this, the importance of further investigation work will be outlined in circumstances that require opening up works or more specialist input. Finally, the various stages of managing the remedial work process will be described, alongside the importance of client consultation of the different options and preferred solutions. Examples of different scenarios leading up to the agreement of the agreed remedial works will be provided.

## 2.2   A Background to Conducting Building Surveys

Carrying out building surveys calls for a comprehensive knowledge of building typologies, together with an understanding of many different types of construction technologies alongside experience in the surveying process. Furthermore, they require the forensic identification, collection and recording of evidence to support a data-led approach to defect diagnosis and remedial works. In this pursuit it is vitally important for building surveyors to assess all the possible causes of defects and not simply jump to assumptions. For instance, there may be multiple causes of defects in some cases, and it would be easy to determine that dampness in external walling is being caused by a visible problem, such as a leaking gutter, but could transpire to be a more complicated issue including rising dampness and water penetration through a defective cavity tray above a window or door opening. In other cases, relatively straightforward building problems could be misconstrued for more complicated and expensive building defects. If a building defect diagnosis is incorrect it is almost certain that remedial works to address the problem will also be wrong, resulting in potentially unnecessary expensive repairs which will have little or no benefit in rectifying the problem. An example could be condensation on internal faces of external walls being diagnosed wrongly as water penetration through the walling structure. In this case, the remedial works to combat what is thought to be water penetration could involve resealing around all windows and doors and repointing brickwork, when the correct remedial works to address condensation would be to introduce more internal ventilation and possibly improve the insulative properties of external walls. Other examples of misdiagnosis of defects could quite easily be made where shrinkage and/or crazing of external wall render or internal plaster finishes are mistaken for failures in the structural integrity of buildings associated with movement cracking from subsidence and buckling or deflection of external walls. Such cases in the past have resulted in legal claims from building owners or leaseholders for professional negligence and damages for monies expended. For this reason, all surveying companies should maintain a minimum level of professional indemnity cover to ensure they are fully covered and protected in these instances. Where there is a potential case of uncertainty as to the source or diagnosis of a building defect, then where appropriate the surveyor should consult with colleagues or specialists to gain a second opinion, in the same way as a medical practitioner would do before diagnosing the source of an illness or disease.

Before surveyors agree to take on a commission to undertake a building survey, they should fully understand what they believe their clients require and the purpose of the survey. There are many different types of surveys that can be carried out on buildings and each one will very much depend on the purpose of what is required from the survey.

If clients are not clear which type of survey they require, then surveyors should outline their options and the respective approximate costs of each and make recommendations where necessary. For instance, condition surveys may be prepared at the start of leases as a record of the condition and state of repair of buildings and used as a reference therefore in landlord and tenant agreements. At the end of leases, a schedule of dilapidations survey will be conducted with reference to the condition survey which will outline the wear and tear, deterioration and defects which have emerged during the term of the lease and set down remedial works to address these. It is normal for schedules of dilapidations to include budget costs for the remedial works and the tenant would be required to pay for these works or instigate the works themselves. Landlord and tenant deliberations on the extent and cost of dilapidations works can become a source of debate and sometimes dispute between the parties, and it is not uncommon for legal action to result from this.

## 2.3 The Process of Undertaking Building Survey

When purchasing a property, it is normal for the purchaser to commission building surveys to assess the condition of buildings. The most proficient survey is normally what is referred to as a 'full structural survey'. These are normally detailed reports and give a thorough description of the condition of all building structures, elements and components. They will normally also give a summary of any necessary remedial works to address defects or items of disrepair. The costs for undertaking these remedial works will normally be a source of negotiation between the buyer and the seller, and sometimes a reduction in price will be agreed between the parties to take account of the necessary works. When surveying dwellings in the UK, structural surveys are not the only type of survey that can be carried out, owing to the substantial costs of undertaking these. It is probably more common for 'home buyer surveys' or 'valuation surveys' to be undertaken. Home buyer surveys would normally be a high-level and summarised version of a full structural survey and would report on the condition of each integral part of a property and any remedial works. A valuation survey would be a relatively short and condensed report on the overall condition of a property and the give a value on the property for mortgage purposes. Valuation surveys normally comprise the minimum requirements that banks and building societies require in order to grant a mortgage on a property. Prior to agreeing to undertake surveys and making arrangements for the inspection, a fee quotation setting out the scope and extent of the survey with any limitations should be provided in writing to clients and adopting a format in line with the Royal Institution of Chartered Surveyors' format and inclusions. The fee quotation should clearly make reference to any services outside the remit of the survey and contain a list of exclusions accordingly. It should also set down the provisions with regard to employer and indemnity insurance held by the surveying company and a complaints procedure in the unfortunate event of a dispute. It is important for a contract to be entered into on the basis of this fee quotation prior to surveys being carried out. It is also important for surveyors to agree with their clients the extent of opening up and making good the access into and around the premises in order to conduct the inspections.

All surveys will be subjected to a minimum level of due diligence in terms of covering all the building elements and sections included in the survey report and in the UK, they normally follow the Royal Institution of Chartered Surveyors format for presenting such reports. Most building survey reports will exclude certain items which are considered either too specialist for the type of survey being carried out and therein beyond the scope of the survey. There may be areas of the building that are regarded as being inaccessible, e.g., underground drainage and manholes, and this is another reason which can lead to them being excluded. There will be certain instances where a particular issue requires the services of another potentially more specialist consultant. For instance, if dampness is detected in external cavity walling there may be recommendations related to further investigatory work using borescope equipment. The borescope would be able to view inside cavities of external walls and assess whether there is any 'bridging' between the outer and inner masonry leaves, possibly caused by mortar droplets on cavity walls, which could be resulting in damp penetration across the cavity. A photograph of a borescope is shown in Figure 2.1. In addition to the above, there may be specialist reports carried out on a particular element of concern on a building. For instance, if a building is showing signs of differentiation settlement or subsidence then a structural engineers report may be commissioned on that particular problem. Likewise, a specialist survey may be undertaken solely on a roof where water penetration through a roof covering is being experienced. Both these types of specialist survey would seek to identifying the cause and scope of the defects and thereafter recommend specialist remedial repair or replacement works to address the defects.

In addition to borescopes, thermal imaging apparatus could also be useful and these normally deploy infrared cameras to detect 'cold bridges' through structures, sometimes associated with water penetration and depletion of insulative materials within building elements, e.g., cavity wall insulation. Thermal equipment can also be useful in identifying parts of buildings that might be causing an excessive degree of heat loss, thereby identifying measures to improve energy efficiency and environmental sustainability.

**Figure 2.1** Photograph of a borescope for inspecting cavities with external walls.

**Figure 2.2** Photograph of a thermal imaging camera commonly used in building surveying.

A photograph of one type of thermal imaging camera commonly used in building surveying is shown in Figure 2.2.

Building surveys may also be linked with maintenance and preparing a planned preventative maintenance programme. Such programmes will predict what level of maintenance needs to be undertaken on a year-by-year basis to keep a building in good repair. Programmes of this nature seek to ensure that building issues do not evolve into building defects and cause ancillary damage to the structure and fabric of buildings. Planned preventative maintenance programmes and work undertaken in conjunction with them are sometimes carried out by specialist facilities management companies who are contracted by client bodies to take the burden of maintenance and facility management away from them.

Undertaking any type of building surveys calls for a forensic, methodical and systematic approach to the inspection process. This is especially the case during the process of defects diagnosis where the surveyor's skills, knowledge, expertise and experience will be called on to providing evidence-based findings, conclusions and recommendations. Desktop investigations prior to arriving at the property can sometimes be useful in identifying local problems in a particular area and this could include flooding and any underground mines. Information of this nature can then be referred to in the inspection reports and can also provide a reference point for closely assessing any problems with the building structures or fabric curtailing from such potentially local conditions. In the case of an area synonymous with mining, careful examination of brick walling could reveal diagonal cracking associated with differential movement and, in more severe cases, subsidence. Likewise, a history of local flooding could reveal signs of previous damp penetration at low level in floors and walls. It is also useful wherever possible to request information, such as plans and elevations of the property, from the building owner in advance and any subsequent improvement or modifications that have been made to the property over the years. It is also worth checking with the building owner or occupier that there are no hazardous materials that the surveyor is likely to come into contact with during the survey process, e.g., asbestos. Other desktop information that may be useful could include previous building surveys of properties. In this way, surveyors could review previous problems with buildings and ascertain whether these have been adequately and competently remedied in the past. If problems have not been addressed, defects may have led to further problems that surveyors will need to look out for on site.

Upon arrival at a property, every surveyor will have their own ways and means of carrying out building surveys. It is wise for any surveyor prior to starting to write the report to have a good walk around the building, both internally and externally. This is to assess the general overall condition of a building and to cross-check whether an external defect is causing internal problems and vice versa. It is also important on this initial walk around to assess any difficulties in carrying out the survey, such as access problems associated with heights, depths and openings, confined areas or unsafe areas of the building. For health and safety reasons and to ensure that the employers' insurance cover is not being compromised, it is usual to carry out a very brief risk assessment and survey plan to confirm that by inspecting the property is not putting the surveyor in any imminent dangers. In carrying out health and safety risk assessments, surveyors should consider lone working, potential hazardous materials, any unsafe debris, unlit areas, uneven surfaces which could lead to trips and falls and any mechanical or electrical hazards e.g., exposed electrical wiring. These can then be referred to in the inspection report. To ensure surveyors do not miss areas and elements of the building that they are inspective a methodical system could comprise inspecting the external areas first and from a top-down approach, working down to ground level. Sometimes high areas of building, for instances, roof areas are difficult to see from ground level and could require good quality high magnification binoculars, such as the ones shown in Figure 2.3, to focus in on certain areas e.g., roof ridge tiles. Other items of equipment may include a damp meter, commonly referred to as a protimeter, for checking for moisture content in floors, wall and joinery (see Figure 2.4).

**Figure 2.3** Photograph of good quality high magnification binoculars.

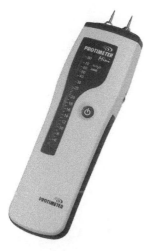

**Figure 2.4** Photograph of damp meter commonly referred to as a protimeter for checking for moisture content in floors, wall and joinery.

Other equipment that surveyors could have in their kit bags could include overalls and face masks for accessing potentially dirty areas such as roof spaces and cellars, boots, crowbars and/or manhole keys for accessing manholes, knives for checking for timber decay and a camera for taking photography of buildings and recording defects and torches for accessing dark areas, such as lofts and under suspended floors. Sometime a tape measure can be useful for recording measurement, e.g., cracking in brickwork. Other equipment could include a set of telescopic ladders for accessing high areas of the building, e.g., flat roof areas, if it is safe to do so and these are illustrated in Figure 2.5. In some cases, ladders may not be adequate or safe for access to very high areas. In such instances, surveyors should agree with their clients for the hiring of access equipment or plant, which could include tower scaffolding, or an access platform lift. Drones may be a more economical and safer alternative where conventional access systems are not financially viable or where there are specific known hazards. Sometimes dyes can be kept in surveyor kit bags for tracking drainage runs and to ascertain the source of roof leaks. In addition, levelling and alignment equipment such as plumb lines, spirit levels and lasers can be useful to check whether the walls or roofs are deflecting and out of plumb.

Sometimes building surveys are undertaken in two stages, with the first stage comprising of an initial inspection to get an overall high-level perspective on a particular defect, and to assess further investigatory works. The second stage would be more in depth and could include carrying out specialist investigatory works, especially in hard-to-access areas or those areas unable to be viewed normally. For instance, if isolated subsidence is noted in one particular area of a building close to an adjoining drain, a camera drainage survey may be required to ascertain the condition of the drainage runs. Water leakage from collapsed or fractured drainage may in this way be revealed to be the underlying cause of the problem, and a drainage survey can accurately pinpoint the exact location of damage and therein the scope of the required remedial works. In some cases, the second

**Figure 2.5**   Set of telescopic ladders for accessing high areas of the building.

stage may consist of lab reports, curtailing with a potential defect. Another example of a second-stage process could be concrete testing following visible failure of concrete structures identified in the first-stage inspection. Tests may be undertaken for the presence of High Alumina Cement (HAC) which, since its introduction in the 1960s, has become a major latent defect, sometimes causing catastrophic structural failures, especially in areas with high humidity levels, such as roofs and walls of areas containing swimming pools. The modus operandi of this second-stage process is to gather enough information on the potential building defect to diagnose its cause and thereafter to be confident that remedial works can be competently designed and specified. In certain cases, during the second stages there may be a need to monitor remedial works and potential defects to see if stabilisation of problems has been achieved or if further problems exist – this monitoring period may be needed across a medium to long period of time. An example would be in the case of movement cracking in external walling due to subsidence. Once remedial works to alleviate the source of settlement/subsidence have been completed it is important to verify whether any further movement is continuing. In such cases 'telltale' plates, which record whether existing walling cracks are stable or opening up, can be installed across the length of cracking and these can be revisited regularly, and data collected as to the position of any further movement.

The exact way in which surveyors choose to conduct their surveys and record their inspection finding varies from one individual to another. Some surveyors choose to use a checklist of internal and external elements and write down their findings in preparation for writing up their reports at a later stage. Others choose to dictate their observations via a dictaphone as they proceed, which an audio typist can incorporate into a written report. Technological advancements have now allowed for handheld devices to be used to record defects and the general condition of buildings. Using software on such devices allows data to be downloaded at a later stage and automatically populate a survey report with limited need for typing up, offering practical and resource advantages in terms of time and cost.

## 2.4   Challenges and Obstacles When Undertaking Building Surveys

Every survey presents different pressures and challenges in the way it is planned and executed. Carrying out inspections in premises which are occupied and accessing all areas around live operations, whilst trying to avoid disruption to end users, can be particularly challenging and requires a careful degree of planning and the help of the owners or occupiers to negotiate around certain conditions and constraints. Conversely, different types of challenges can present themselves when organising inspections of vacant, potentially derelict buildings, where lone working may be needed and it is necessary to ensure that a safe survey plan is established. Working in schools with requirements for safeguarding children may involve obtaining DBS (Disclosure and Barring Service) checks or alternatively surveyors being escorted around at all times. Schools, colleges and universities during term time can present problems for building inspections, with special arrangements being required to access specialist, potentially hazardous areas such as laboratories and planning to access classrooms around timetabled teaching sessions. Hotels, retail and commercial premises can present similar challenges to being confident that all areas can be accessed without unduly affecting and disrupting the end users. All types of property and categories of survey can present specific problems where intrusive and/or destructive investigations are required. This could involve opening up walls or floors to gain visibility or access for further investigations. Clearly, such works in their own right will create disruption, so alongside all making good works, these are best planned out of normal working hours if at all possible. The most important thing to consider as part of the building surveying process therefore, taking account of all the possible necessary arrangements, is to maintain excellent communications with clients throughout to manage their expectations, especially since abnormal arrangements and investigatory works could add significant additional costs to the survey work. In such cases, clients may have to apply for additional funds to bolster their budgets prior to works proceeding.

Challenges when conducting building surveys could also emanate from establishing accurate measurements to assess the condition of building elements. Having apparatus that can be relied on to measure the moisture content of walls, floors and internal joinery can be an essential facet in assessing whether dampness is causing problems within buildings and providing the right conditions for mould growth, deterioration of building fabric, and for rot to thrive. Even with the right instruments calibrated to an accurate standard of measurement, surveyors need to be careful when interpreting the results. One example could be that occasionally damp meters, manufactured to measure electrical resistance through fabric, cannot distinguish between soluble salts in building materials such as plaster, and the actual moisture content of the actual materials that are tested. Accordingly, this once again could lead to a wrong diagnosis and expensive unnecessary remedial works being recommended to clients.

When carrying out building inspections it is important wherever possible for surveyors to define the severity of defects against predetermined metrics to assist them in making their assessment. This is to avoid cases where one surveyor regards a particular building problem as moderately severe, and another classifies the same problem as very severe. This is particularly critical if a team of surveyors are carrying out a programme of 'stock' building inspections for a client, e.g., a housing association, when consistency is required across many different building reports. By way of an example, in the case of movement cracking in brickwork walls, classifications could be agreed as to a scale or range of thicknesses and lengths of visible cracking, where '1' could denote less severe hairline cracking below 2 mm, and '5' could denote more than 15 mm, which could be categorised as very severe. By utilising this methodology, building surveyors can build up a picture of defects on a particular property which might all be interrelated to a common cause or symptom. Using this method, it is also critical to record the configuration of cracking, as this is normally a telltale sign of what is causing the cracking. Vertical cracking of external masonry walls predominantly at corners of buildings (but not always) are normally associated with thermal movement of the structure. Such cases are usually attributable to design deficiencies where movement joints have not been adequately designed, specified and built into the structure. Alternatively, diagonal cracking could be associated with differential settlement of the walling possibly due to foundation failures. In severe cases, subsidence could cause differential settlement across many different areas of a building, resulting in varying degrees of cracking across many locations. It is not always possible to monitor this problem by visual means; instruments for measuring cracking or dampness may be needed. In the former case, micrometers can normally give accurate dimensions (see Figure 2.6) over time but sometimes tell-tales (see Figure 2.7) can be fitted across the cracks to measure movement.

Another example where categorisation is important could be water penetration into buildings which could be causing extensive dampness to some areas of the building and limited moisture build-up in others. Being able to assign predetermined degrees of severity to each different area of dampness from minor to very severe is again a way of introducing consistency into the survey reports. In this case, it may also be necessary to monitor the movement or dampness over a period of time to assess whether the defect is progressing, or following remedial works, if the problem is now fully addressed and thereby historical. To assist in this pursuit, damp meters can accurately record the moisture content of structures, and photographic evidence can record the degree of spread and exposure.

In all cases, the above methods may only give surveyors an insight into whether the defects are still prevalent, but not give enough information for diagnosis of the precise cause to allow remedial works to be specified. This dilemma may be common where there are potential defects in walls that are not visible. In such cases, opening up works may be required in order to inspect further into the structure. This is common on external cavity walls where cases of dampness or movement could be being caused by defects within the cavity area. Dampness could, for instance, be emanating from

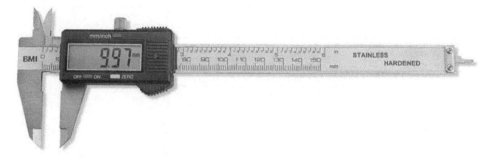

**Figure 2.6**  Photograph of a micrometer.

**Figure 2.7**  Photograph of a telltale.

bridging of the cavities caused by mortar deposits between the inner and outer leaves of brick and blockwork, whereas movement cracking could be resulting from corrosion of cavity ties or lintels. Normally this type of opening up works can be undertaken by a small builder who can sometimes reinstate the building structures and fabric back to their original form once the intrusive investigations have been properly carried out, recorded, costed, planned and managed. Examples of opening up for further intrusive investigation are also synonymous with the following cases, as shown in Figure 2.8.

## 2.5  The Importance of Building Investigations

When surveyors are undertaking building surveys, as mentioned earlier, it is essential for them to adopt logical and pathological methodologies. This can include the sequence of the inspections and the mode by which they record their findings. In undertaking

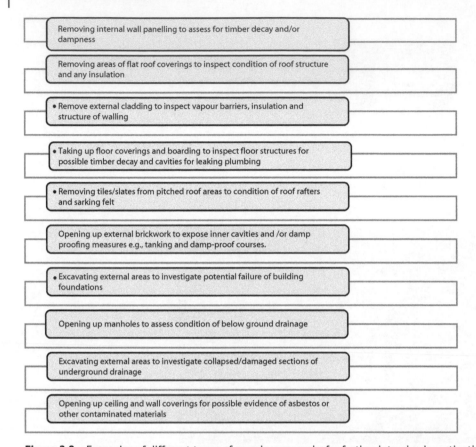

Removing internal wall panelling to assess for timber decay and/or dampness

Removing areas of flat roof coverings to inspect condition of roof structure and any insulation

• Remove external cladding to inspect vapour barriers, insulation and structure of walling

• Taking up floor coverings and boarding to inspect floor structures for possible timber decay and cavities for leaking plumbing

• Removing tiles/slates from pitched roof areas to condition of roof rafters and sarking felt

Opening up external brickwork to expose inner cavities and /or damp proofing measures e.g., tanking and damp-proof courses.

• Excavating external areas to investigate potential failure of building foundations

Opening up manholes to assess condition of below ground drainage

Excavating external areas to investigate collapsed/damaged sections of underground drainage

Opening up ceiling and wall coverings for possible evidence of asbestos or other contaminated materials

**Figure 2.8** Examples of different types of opening up works for further intrusive investigations.

diagnosis of building defects each potential defect should be considered carefully to forensically assess all the symptoms that have been presented and to examine all the possible or probable sources and causes of the problems. This process calls for systematic elimination of possible causes, as usually the sources of the problems are not always initially obvious. It can involve the testing of different hypotheses relating to symptoms and possible causes. For instance, structural cracking could be resulting from expansion associated with thermal moisture forces, or conversely, gravitational forces associated with foundation failures and/or ground subsidence. This process may lead surveyors to not just one possible cause but to a narrowing down of potential causes for further reflection. In this way, the true cause of the defect can hopefully be identified, avoiding the risk of false diagnosis leading to remedial works that are ineffective and wasteful. The successful diagnosis will ultimately allow a technical report to be prepared which will detail the extent of the problem, how the underlying causes have been arrived at and, most importantly, recommendations of all the necessary remedial works. Sometime the costs and timescales associated with the repair works can be outlined, especially if the surveyor is being appointed by the client to oversee the programme of remedial work. These technical reports will normally include details relating to insurances, liabilities and quality compliance. They will sometimes also include logistics for implementing the repairs in occupied premises where there is disruption to end users, and health and safety are paramount. They could also offer recommendations or names of specialist building contractors that would be appropriate, and offer the necessary technical expertise to undertake all or some of the remedial works. In certain instances, there may be waivers in the technical report that call on more specialist reports from manufacturers, or installers of specialist cements and components, to be consulted or employed. This could be the case where there are failures in specialised roof systems and where only the original roofing manufacturer that installed is suitably qualified to complete the works to the required standard and provide a guarantee to underwrite this commitment. Other examples could relate to curtain walling systems or specialist areas of mechanical and electrical installations and plant.

## 2.6 Managing the Remedial Work Process

As previously referred to, the technical report will normally outline different recommendations associated with remedial works to address different defects. In some cases, for each item of remedial works there will not be one option but many different options which will give clients the choice in terms of affordability, priority, practicality and possible disruption to the building occupiers. Sometimes each option can relate to temporary repairs to monitor the problems, or to the availability of client's funds, or the 'full and final' solution which will seek to cure the problem permanently, albeit at a higher level of cost. In addition, there may be intermediary options which provide a 'halfway house' in addressing the building defects. The following scenario details one example.

The above example gives two far-ranging and diametrically opposed options to addressing the problem, depending on the constraints of the buildings, future plans and affordability. Notwithstanding this, there may be more intermediary measures between these two opposing solutions. There will be a range of works which could include a large degree of maintaining asbestos in its current state, inspecting, recording and managing as required,

---

**Asbestos contamination within a building**

A technical building survey has revealed that there is extensive and potentially dangerous asbestos within an occupied building. Taking into account that the asbestos is widespread throughout the whole building and the building has many staff working within it, the client has asked the building surveyor who is overseeing the remedial works to present a report on the options available to them. The following options have been outlined:

**Record the full extent of the asbestos on a risk register and monitor for any damage**

The asbestos is currently not in a state where it is an immediate threat to the health and safety of the building occupants. Some of the asbestos is of the fibrous type, hidden behind wallings and other areas containing the amosite asbestos cement category which forms part of ceilings and wall panelling. All the visible asbestos is largely intact and not currently regarded as being in a condition which would cause release of fibres into the air, thereby causing infection to building occupiers. Owing to the availability of funds for part or complete removal, this scenario of recording the full extent of the asbestos on a risk register and monitoring for any damage would seem to present the best alternative. This position is predicated upon the alternative, whereby removal works could be both expensive, unaffordable and hugely disruptive to the building and business. Recording and monitoring the state of the asbestos could also be a suitable alternative where there are longer term plans to demolish and redevelop the building, whereupon the asbestos could be more affordably and safely removed in one stage once the building is vacated. With this alternative, it is important to make clients aware that it requires continual inspection and recording of the condition of the asbestos on an asbestos register, as any interference or disruption to it could render it to be hazardous. Health and safety procedure will need to be put into place to record inspections, and building maintenance operations will have to stringently adhere to strict protocol when being carried out in close proximity to affected areas of the building. In some cases, this may prove onerous for building owners, preventing them from carrying out interventions and works of alteration and improvement.

**Undertake remedial works to remove all traces of asbestos**

Where the asbestos is in a particularly dangerous state in certain areas and/or where clients want full freedom to undertake improvement, alterations, or refurbishment works unhindered by the presence of asbestos, removal may present the best alternative. In this scenario, it should be recognised that if the asbestos is extensive and requires intrusive/destructive opening up works to access it, then it can be very expensive to remove. In addition, it is likely that the removal works will have to be carefully managed and take place in phases, covering particular areas for removal at one time, and those areas to be accessed for the works must be safely isolated from the remainder of the building. This clearly will call for occupants to be decanted to other areas as the works progress, and this can be hugely disruptive and disconcerting for them.

but with a programme put in place to periodically remove those areas deemed to present a more imminent and important threat. Organisations like to adopt this hybrid approach to avoid too much immediate disruption and expense in favour of a managed rolling programme of removal as part of their respective long-term maintenance planning.

Similar scenarios for other building problems may exist for programmes associated with roof, window, door, curtain walling or replacement, where a more phased, programmed approach may be more advantageous rather than carrying out the works as one project. Mechanical and electrical replacement could also be managed in this way to maintain levels of heating and servicing in buildings at all times.

## 2.7 Summary

Conducting building surveys calls for the forensic identification, collection and recording of evidence to support a data-led approach to defect diagnosis and a comprehensive knowledge of building typologies. In this pursuit it is vitally important for building surveyors to assess all the possible causes of defects and not simply make assumptions, as relatively straightforward building problems can sometimes be misconstrued for more complicated and expensive building defects, and vice versa. If a building defect diagnosis is incorrect, it is almost certain that remedial works to address the problem will also be wrong, resulting in potentially unnecessary expensive repairs which will have little or no benefit in rectifying the problem.

There are many different types of surveys that can be carried out on buildings and each one will very much depend on the purpose of what is required from the survey. If clients are not clear which type of survey they require then surveyors should outline their options, the respective approximate costs of each, and make recommendations where necessary. The most proficient surveys are are 'full structural surveys'. These are normally detailed reports and give a thorough description of the condition of all building structures, elements and components. It is, however, probably more common for 'homebuyer surveys' or 'valuation surveys' to be undertaken for domestic properties in the UK. A homebuyer survey would normally be a high-level and summarised version of a full structural survey and would report on the condition of each integral part of a property and any remedial works. Conversely, a valuation survey comprises a relatively short and condensed report on the overall condition of a property, and give a value on the property for mortgage purposes. It is important for a contract to be entered into on the basis of a fee quotation prior to a survey being carried out, and for surveyors to agree with their clients the extent of opening up and making good, and access into and around the premises in order to conduct the inspections. Most building survey reports will exclude certain items which are considered as either too specialist for the type of survey being carried out and therefore beyond the scope of the survey. This could include those areas that cannot be readily inspected, e.g., underground drainage, or those elements which require further specialist advice, e.g., building services installations.

Building surveys may also be linked with maintenance and preparing a planned preventative maintenance programme. Such programmes will predict what level of maintenance needs to be undertaken on a year-by-year basis to keep a building in good repair. Programmes of this nature seek to ensure that building issues do not evolve into building defects and cause ancillary damage to the structure and fabric of buildings.

Undertaking any type of building survey calls for a forensic, methodical and systematic approach to the inspection process. Furthermore, desktop investigations prior to arriving at the property can sometimes be useful in identifying local problems in a particular area, and this could include flooding and any underground mines. Other desktop information that may be useful could include previous building surveys of properties. Sometimes building surveys are undertaken in two stages, with the first stage comprising an initial inspection to get an overall high-level perspective on a particular defect and to assess further investigatory works. The second stage would be more in depth and could include carrying out specialist investigatory works especially those hard-to-access areas or unable to be viewed normally.

On arriving at a property, the surveyor conducts an initial walk around to assess any difficulties in carrying out the survey, such as access problems associated with heights, depths and openings, confined areas or unsafe areas of the building. For health and safety reasons and to ensure that the employer's insurance cover is not being compromised, it is usual to carry out a very brief risk assessment and survey plan to confirm that inspecting the property is not putting the surveyor in any imminent danger. The exact way that surveyors choose to conduct their surveys and record their inspection finding varies from one individual to another. Some surveyors choose to use a checklist of internal and external elements and write down their findings in preparation for typing up their reports at a later stage.

There are many challenges and obstacles when undertaking building surveys. Carrying out inspections in premises which are occupied and accessing all areas around live operations whilst trying to avoid disruption to end users can be particularly challenging and requires a careful degree of planning and the help of the owners or occupiers to negotiate around certain conditions and constraints. All types of property and categories of survey can present specific problems where intrusive and/or destructive investigations are required. This could involve opening up walls or floors to gain visibility or access for further investigations. Clearly, such works in their own right will create disruption, so all making good works are best planned out of normal working hours if at all possible. When carrying out building inspections it is important wherever possible for surveyors to define the severity of defects against predetermined metrics to assist them in making their assessment. This is to avoid cases where one surveyor regards a particular building problem as moderately severe, and another classifies the same problem as very severe. Opening up works may be required to inspect further into the structure and fabric of buildings, and this can sometimes include taking up floorboarding, uncovering roof timbers, removing wall panelling and exposing ceiling and roof voids.

When surveyors are undertaking building surveys it is essential for them to adopt logical and pathological methodologies and this can include the sequence of the inspections and the mode by which they record their findings. In undertaking diagnosis of building defects, each potential defect should be considered carefully to forensically assess all the symptoms that have been presented and to examine all the possible or probable sources and causes of the problems. This process may lead surveyors to not just one possible cause but to a narrowing down of a number of potential causes for further reflection. Thereafter, the successful diagnosis will ultimately allow a technical report to be prepared which will detail the extent of the problem, how the underlying causes have been arrived at and, most importantly, recommendations for all the necessary remedial works. Sometimes options for different remedial works will be given. These can relate to temporary repairs to monitor the problems or owing to availability of client's funds or the 'full and final' solution, which will seek to cure the problem permanently, albeit at a higher level of cost.

# 3

# Common Maintenance Issues and Managing Defects

## 3.1   Introduction

This chapter is focused on the pathology of buildings with particular emphasis on building degradation, and will describe the ways and means of identifying and addressing potential defective elements and defects within buildings. In this pursuit, it will examine the different methodologies for diagnosis of defects and their root cause (rather than just the symptoms of defects) to assist in identifying the most appropriate type of repair.

An analysis of the underlying conditions and mechanisms which lead to defects and building failures as a result of poor design and quality of build during the construction stage or refurbishment will be covered, from dampness to structural movement, corrosion and timber decay. Furthermore, the role of maintenance across the life cycles of buildings will then be discussed and the influence that this has on their preservation and deterioration. The debate around this important area will then extend into the examination of planned and reactive policies for maintenance management of buildings estates.

The challenges associated with identifying the underlying causes of building failures will be presented, where one or more possible different diagnoses may be initially debated. In this context, the requirements for further investigation will be examined in determining the underlying root cause of problems more accurately and confidently. The different categories of building failures which lead to the emergence of defects will then be discussed together with the mitigation measures that should be undertaken to reduce the risk of such failures. In addition, the individual forces at work leading to different types of defects, and how these can vary depending on climate and the geographical locations around the world, will be articulated. Furthermore, the difficulties and challenges of correctly identifying faults and defects to building services such as heating, lighting, fire alarms, cooling and other forms of gas and electrical installation will be explained, and practical examples provided.

The importance of understanding the nature and effect of agents that can lead to building defects will be examined in detail, including human misuse of buildings and deliberate vandalism, together with electromagnetic, mechanical, chemical, building user, and thermal and biological-related forces and effects at work. The wrong use of materials and components and design defects will be discussed, and how these can develop into latent

*Introduction to Built Asset Management*, First Edition. Dr Anthony Higham, Dr Jason Challender, and Dr Greg Watts.
© 2022 John Wiley & Sons Ltd. Published 2022 by John Wiley & Sons Ltd.

construction defects in the future, therein introducing maintenance liabilities for building owners. The magnitude of managing remedial works and the associated disruption that this can bring for building users will then be investigated, with reference to different examples in practice. Dilemmas associated with repair or renewal decisions will be introduced and debated in detail and the overall effects that decision making can have on future maintenance and life cycle costings analysis. This will entail a discussion around capital cost (CAPEX) versus operational cost (OPEX) perspectives. Following on from this, the various influences of different factors on the repair or renewal decision making will be covered, including capital funding, building longevity and potential disruption from proposed remedial works.

Managing the remedial works processes to address maintenance issues and defects will be examined with a focus on best practice and achieving the correct level of supervision and quality control throughout the contract period. This will introduce quality control and best practices in selecting the right design, team and contractors from the perspective of attaining the necessary skills, experience and competency levels required to undertake projects successfully. Finally, measures to mitigate and prevent defects will be covered, and the importance of allowing access provisions in buildings to enable and carry out regular maintenance inspections as part of planned preventative maintenance programmes.

## 3.2    Exploring the Pathology of Building Maintenance Issues

The pathology of building maintenance issues revolves around the investigation, inspection and reporting of building defects. According to Thomas (2013), building pathology is defined as 'the systematic treatment of building defects, its causes, its consequences and its remedies'. The principal aims and objectives around building pathology are related to the need to comprehend the processes of building degradation. In this way, tools and methodologies can be introduced to identify potential defective elements and defects within buildings. This should facilitate solutions to be devised to eliminate or decrease their detrimental effects, thereby avoiding unnecessary or unforeseen maintenance expenditure.

Building pathology is related to the school of applied sciences and for this reason it is universally regarded as a specialist area in its own right. It is founded on investigation work to identify the source or sources of building problems. Thereafter, it is concerned with diagnosis of the underlying technical issues which could be causing the problems and then determining the extent and specification of the required remedial works. The findings from this investigation and diagnosis work should identify specific components, elements or areas of a building where problems are arising and the underlying defects or malfunctions that are responsible for them. In this regard such building problems may relate to individual building systems or alternatively groups of systems. For instance, water penetration into a building emanating from a flat roof covering could be caused by a single tear in the roof covering (single building system) or a combination of different envelope defects, including the roof and walls (group of systems).

The term 'pathology' has only been applied to buildings relatively recently and was previously only used in conjunction with the health sector when describing medical problems in people and diagnosing types of conditions and diseases. In this context it allowed medical practitioners an understanding of particular health complications and to recommend a course of medicine or procedure. Buildings and people, in terms of pathology, have many aspects in common. The condition of a building is not always correlated with its age, on the basis that it is possible for a newer building to have more defects and signs of wear and tear than an older building, as a result of poor design, workmanship and quality of build. A building's poor condition could also be a result of neglect and a lack of routine maintenance, in the same way that a person's health is not always related to how old they are but is heavily influenced by lifestyle, exercise, diet and maintenance of their well-being.

Normally, when considering the process of building pathology, it will require suitably qualified specialists to undertake inspections and building surveys. Building pathologists will try to identify, assess, eliminate or control (through a managed process) building defects and their causes. In the UK, building pathologists are normally building surveyors and members of the Royal Institution of Chartered Surveyors (RICS). However, if a particular building defect is related to major complicated structural problems, then a consultant structural engineer and possible member of the Institute of Structural Engineers (IstructE) may be appointed. Normally these structural engineering professionals are more specialised in dealing with structural issues such as these and can more proficiently prescribe the most appropriate remedial works. In a similar way, if a building defect is related to mechanical and electrical issues, possibly related to heating, ventilation or cooling problems, then a suitably qualified mechanical and electrical engineer may be appointed. These individuals are normally members of the Chartered Institute of Building Services Engineers (CIBSE) and are more specialised in dealing with issues such as these and can more proficiently prescribe remedial works to building services.

## 3.3 Context to the Discussion on Building Defects

A building defect could be described as a fault, deficiency or imperfection in a building which causes it to not properly function in terms of appearance, performance or usage. When building surveyors are determining remedial works, following diagnosis of defects, it is important for them to focus on the root (cause) of the problem rather than simply address the symptoms of the defect. An example is the structural movement cracking of an external wall, which requires structural foundation remedial works to address differential movement and subsidence, rather than simply repointing cracked mortar joints. This should ensure that repairs are not temporary superficial 'fixes' but, conversely, present long-term solutions.

When investigating, diagnosing and specifying remedial works to address defects, building surveyors should be mindful through their knowledge, experience and training of the underlying conditions and mechanisms which lead to defects and building failures. Certain building failures result from poor design and quality of build during the construction stage or refurbishment. Other defects may simply arise due to natural forces associated with weathering and normal use. In the latter case, examples could include weathering of timber cladding that has been open to the elements for many years, causing shrinkage cracking and rot in certain instances. Clearly, the degree of maintenance that buildings receive over their life cycles will strongly influence the degree to which they are preserved or deteriorate. Planned maintenance policies in this regard aim to preserve buildings to a perpetually good standard and involve routine frequent interventions. Where maintenance policies are more reactive in nature, this is where problems can be exacerbated. For instance, there may be water penetration through a roof covering that goes unchecked for a number of months or years owing to a lack of planned maintenance, and reactive maintenance only occurs once the problem becomes apparent. In such cases, hidden water damage may have resulted in dry rot in the roof timbers, which is a much bigger problem to resolve than the initial roof leak.

Diagnosis of building defects is not always straightforward, as there may be more than one force at play in terms of multiple sources of different problems. In such cases, this can sometimes complicate normal lines of investigation and enquiry and present difficulties in determining the cause and associated remedial works. Complex investigations can arise in the following circumstances:

1. Wrongly specified materials and construction detailing. Roof abutment detailing and basement waterproofing is commonplace in this defect category, and it may be many months before water penetration can be first identified.
2. Correct materials are specified but detailed in an unsuitable way for the nature of the product, component or material. By way of an example, the correct roof flashing detailing for lead may be very different to that used with a different roofing material such as bituminous felt.
3. Wrong installation of products, components or material, albeit specified and detailed correctly. For this reason, it is important to not always rely on building contractors to complete installations without a supervising officer, which could be the architect or architectural technologist inspecting to confirm adherence to construction drawings and other design documentation.
4. Wrong assessment of conditions associated with the design, which could include exposure to the weather or structural loading. An example of this could be the specification and design of steelwork that does not meet accepted factors of tolerance and building regulations.

When considering the typology of defects in buildings one should distinguish between different types of problems. According to Addleston (1989, as cited in Thomas 2013) these can be categorised by type and cause into (i) settlement and subsidence; (ii) condensation, dampness and water penetration; (iii) loss of bond and adhesion; (iv) decay and corrosion;

and (v) movement and cracking. These five categories can be broken down into subcategories and these are illustrated in Figures 3.1, 3.2, 3.3, 3.4 and 3.5. Clearly, the extent of each of the defect categories will be largely dependent on the different forms of construction. In the UK, house building tends to be heavily geared to traditional forms of construction with brick and block cavity external walling and pitched tiled roofs. Conversely, in Scandinavia the predominant domestic construction comprises timber framed and cladded buildings of a lightweight, and sometimes modular nature. Alternatively, in some parts of Europe, such as Spain and France, cast concrete structures for housing and commercial buildings is predominantly the norm.

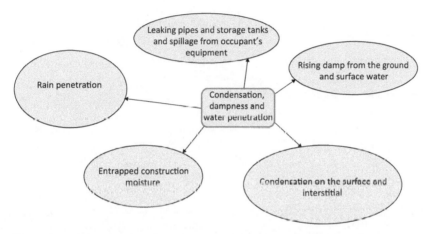

**Figure 3.1** Subcategories of condensation, dampness and water penetration.

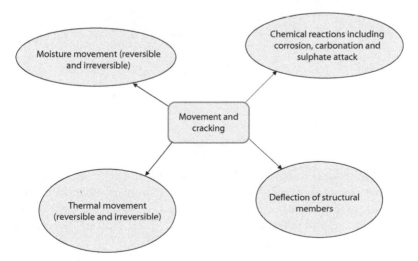

**Figure 3.2** Subcategories of movement and cracking.

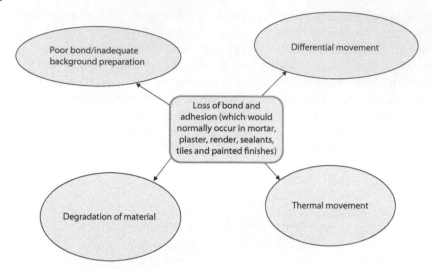

**Figure 3.3** Subcategories of loss of bond and adhesion.

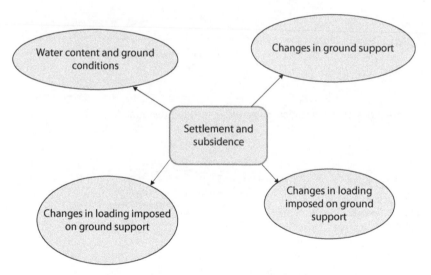

**Figure 3.4** Subcategories of settlement and subsidence.

When we consider building services such as heating, lighting, fire alarms, cooling and other forms of gas and electrical installation, it can be more problematic to categorise defects and faults owing to the specialist nature of technologies in some of these systems. Figure 3.6 does, however, attempt to illustrate some of the types of failures, defects and faults in mechanical and electrical systems.

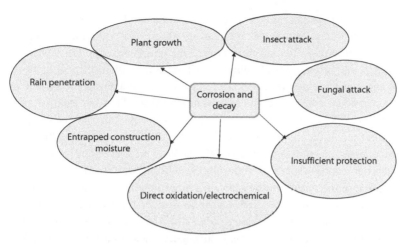

**Figure 3.5** Subcategories of corrosion and decay.

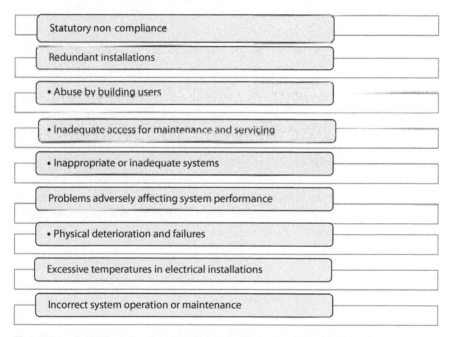

**Figure 3.6** Possible defects and faults in mechanical and electrical systems.

## 3.4 The Importance of Understanding the Nature and Effect of Agents That Can Lead to Building Defects

The causes and agents at work that can introduce some of the defects mentioned can be categorised into different bands and groups and these include electromagnetic, mechanical, chemical, building user, and thermal and biological-related forces and effects.

Examples of building user causes can range between deliberate vandalism to accidental or wrongful misuse. Conversely, an example of chemical agents at work could include oxidation, acid attack and water damage to materials. One of the main causes of defects can be a result of poor design and simply not allowing for thermal or moisture movement in building elements and components. An obvious example could be where long lengths of brick walling are constructed without adequate movement joints. Thermal expansion of structures such as brickwork without provision for movement can cause stresses within these structures, leading to cracking and potential structural failures. Different building materials and components have far-ranging coefficients of expansion when heated. It is for this reason that when components and materials come together to form building elements an understanding of the nature of the materials and how they will react with each other is critical. The example of Liverpool's Catholic Cathedral, known now as the Metropolitan Cathedral of Christ the King, shortly after it was built in the 1980s presented a classic example of where a roof covering was insufficiently designed for the cathedral and detailed without sufficient provision for movement joints. Thermal movement of the lead roofing system was not fully understood, with inadequate measures installed to cater for such movement in the materials. This caused water penetration through the roof covering and ultimately led to its failure, resulting in expensive and long-drawn-out legal proceedings by the Diocese of Liverpool against the architects for the costs of remedial works.

Notwithstanding the above causes of defects including those of design, the most common defects arise from poorly manufactured materials, components and elements alongside poor workmanship in the construction process. It is therefore important wherever possible to only use materials and building components that are adequately accredited for the standards or quality. Normally, to attain such accreditation requires these materials and products to be robustly tested in lab conditions to verify their suitability when faced with potentially damaging climatic and environmental conditions. In the UK, accreditations for materials can come from British Standards whereby an accreditation number is assigned to a particular building product. Accreditation and the use of quality tested components are particularly important for mechanical and electrical installations to ensure that the systems perform to predetermined standards and meet prescribed levels of longevity and durability. To combat poor workmanship in the build process it is essential to only appoint suitably qualified and appropriately selected main contractors, subcontractors and specialist contractors who are suitably qualified for the nature of work they are required to carry out. Every project is unique and bespoke in many ways, with varying degrees of complexity and wide-ranging values. For this reason, it is important to undertake a robust tender evaluation and selection process to ensure that the most suitable contractors are appointed for a particular project, and only use contractors that can demonstrate knowledge, expertise and a successful track record for similar types of work. Checks should be made wherever possible on previous projects that they have undertaken, with references provided by previous clients. This should ensure that a reputable contractor is appointed with all the required levels and degrees of competence. In the selection process the nature of the proposed project should be considered against the skills and expertise of those contractors who are being considered. After all, a contractor that may be suitable to construct a £200k house refurbishment will seldom be appropriate for a £50 m new build and vice versa. In addition to selecting the right contracting team it is also vitally important to have arrangements in place to monitor and inspect works completed to ensure the right level of quality is being achieved. Normally this task would fall to members of the design team, and they should cross reference the works against design drawings and specifications to confirm close adherence with the contract documentation. Any deviation from the design drawings and specification should be reported to the supervising officer on the project, which in normal circumstances would be the client's project manager. The project manager would then inform the contractor of the non-compliant and potentially defective works and request that the matter is addressed.

Another common cause of premature failure and defects is a general lack of understanding and lack of maintenance of buildings. Although most organisations have planned preventative maintenance systems in place which seek to address maintenance issues before they become more serious defects, there are still those who will only action works on a reactive basis. This can lead to minor repairs becoming major repairs and is particularly common in vacant and derelict buildings. It is not unusual, by way of an example, for a simple roof leak in a building that has been left unattended and unchecked for a long period of time to have caused problems associated with dampness and dry rot in roofs and walls, potentially requiring major remedial works. Wherever possible, architects and engineers should design buildings to be as maintenance free as possible. This involves not only specifying robust building products and materials but allowing for access for maintenance. An example could be to allow at the design stage for access gantries for roofs to inspect, repair and clean roof coverings and roof lights. The same safe and easy access could be used to maintain roof-sited mechanical and electrical plant including ventilation and cooling systems. It is when such simple and safe access is not designed into buildings that problems can arise with occupiers and owners not wanting to buy or hire expensive access plant or scaffolding to reach those areas that are hard to get to.

## 3.5 Dilemmas Associated with Repair or Renewal Decisions

Other dilemmas can be introduced when deciding on the right level of remedial works to address defects and maintenance issues. The repair or replace question frequently arises when considering building components and elements and how they interact and work with each other. It is a similar scenario to car maintenance in sometimes considering whether a patch repair will be sufficient over time or merely storing up a bigger problem for a later time. An example could be an old roof covering that is clearly showing obvious signs of wear and tear. The question arises should one continue to arrange patch repairs on the roof covering when water penetration occurs or renew the roof covering entirely. Patch repairing old roofs can sometimes be cumbersome and disrupt other areas of the existing roof farther from the actual roofing operations. This can lead to more costs and disruption being incurred over a longer period of time than simply replacing the roof and curing the problem once and for all. Furthermore, patch repairs on old roofs rarely can be guaranteed, but normally a new roof covering carries a minimum level of 10 years' guarantee on materials and labour which can give building owners some added assurance regarding maintenance costs in the future. Other dilemmas could relate to old mechanical and electrical installations, e.g., gas boilers that could carry substantial maintenance liabilities and future high energy costs associated with inefficient and outdated technologies. A decision might be made in these scenarios to replace such plant with more reliable, cost-effective and carbon-friendly technologies. One factor which could influence the decision to renew plant with more energy-saving technologies, possibly involving renewable technologies, could be the potential for government grant funding. In the UK there have been many schemes such as The Green Grant Fund which was set up to encourage householders to install energy primary and secondary environmentally sustainable measures. These included replacement of single glazed windows and doors for double and triple glazed alternatives, alongside other measures including the installation of cavity wall and loft insulation. In addition, the UK Government announced a £134 m 'Cleantech Funding' in 2020. This was designed as part of the Government's green agenda to meet its zero carbon targets by the prescribed date of 2050, and with the aim of helping to promote clean growth projects, develop new technologies, create new jobs, drive productivity and tackle climate change. Other considerations for the repair or renewal question might be dependent on the historical status of a particular building. If a building is listed, it would be seen to be more acceptable where possible to conserve and repair individual elements of its composition rather than trip out historic fabric and replace with new. Conservation officers would rarely allow whole elements to be replaced without a strong case for doing so and evidence to show that these are beyond repair. The case of timber decay, and particularly dry rot, in the past has necessitated considerable strip out and removal of historic fabric. However, in recent years preservation treatments have improved, which has allowed some infected timbers to be retained and strengthened without risking spread of dry rot to other areas.

Sometimes remedial works do not solely look to repair or renew a particular element back to its former state but to improve and adapt the design to make it function more effectively. One example could be a roof that has limited overhang or insufficiently sized rainwater gutters and downpipes to prevent rainwater cascading down façades, causing dampness issues. In these cases, consideration should be given to undertaking improvement and upgrading works to avoid historical problems becoming future maintenance issues. Other similar cases which seek to address design issues in the original build could include removing substandard materials and components. The Grenfell Tower tragedy in 2010, where flammable cladding was found to have resulted in the building becoming ablaze, leading to the death of 70 people, has become a landmark case on poor design and quality control. Many owners and leaseholders are left with the ongoing problems associated with the cost of replacing inappropriate cladding at what can amount to a significant level of expenditure for retrofitting new non-flammable alternatives.

There are many factors which can influence surveyors' recommendations on the renew or replacement question, but whichever option they select should be arrived at through a carefully considered mechanism and one ideally which the building owner should be consulted on. After all, it would be pointless and counterproductive for a surveyor to recommend a large programme of building replacement if the building owners have plans to relocate or simply cannot afford the expenditure that a large refurbishment/redevelopment would require. Through consultation with the building owners a programme of short-term temporary repairs may serve their purposes better than a large-scale/long-term renewal programme. Accordingly, there is both technical skills and intuition required on the part of surveyors in optimising the choice of the most economic and effective recommended repair/renewal solution. Notwithstanding this premise, it is always the client, which is normally the building owner or their representative, that will make the final choice depending on their future plans, aspiration, resources and budget. They will, however, be influenced by the surveyor's technical report and so it is important that this clearly sets out the different options, and the benefits, disadvantages and implications of each one. In a scenario that the repair or renewal costs are disproportionately high to the value of a building, the client may decide to demolish and redevelop, change the use of the building or simply sell the building and relocate elsewhere.

In certain instances, building owners may choose to do nothing when faced with a major defect and simply tolerate the defect and its effects if not too disruptive or damaging to their business operations. One example could be where a basement cellar is letting in water owing to defective tanking. Any remedial works could be considered to be too expensive and disruptive to justify the benefits of making the basement area watertight. Accordingly, the building owner may allow the ground water to seep through and leave to drain naturally or alternatively pump the water away.

## 3.6 Managing the Remedial Works Process to Address Maintenance Issues and Defects

When the remedial works are agreed between the client and their supervising officer/surveyor overseeing the work, a brief outlining the works should then be prepared with budget costings and a proposed plan of implementation. The implementation plan will normally have a programme of timelines for different stages of the procurement process from start to finish. For instance, there will be a design period whereby tender documents are prepared, including drawings, specifications, and specialist reports. For structural works this will normally entail the preparations of structural calculations for any repairs or reinforcement works which will form part of a building regulations submission. Sometimes quantities will need to be prepared to assist tendering contractors in formulating prices for measured works. The next stage will be to tender the works, and this may involve selecting a handful of preferred contractors who are deemed suitable for the nature and value of the works. A pre-qualification process may take place to formulate a shortlist of suitable contractors; this may consist of each potential contractor completing a questionnaire and providing supporting information for their application to be included on the tender list. The works will normally then be tendered and evaluated on predetermined criteria by a panel of individuals agreed between the client and the supervising officer/surveyor responsible for overseeing the works. Once a contractor is selected, contracts will be signed based on the agreed tender price and a programme of delivery. Contractors will be responsible for providing evidence of all insurances (employer's liability and professional indemnity) alongside all the health and safety documentation including risk assessments and method statements.

The works will be closely managed by the supervising officer, normally the surveyor who is overseeing the works, and this will involve frequent inspections of the works and certification at various stages of the project. It may be necessary to have reviews of the work if there are unforeseen areas of work that cannot be adequately specified owing to access difficulties and/or opening up works via destructive intrusive investigation. In some instances, building contracts for repairs and remedial works can be undertaken in two separate phases. Phase one could consist of works to access areas and provide the means for inspection of elements that would otherwise be inaccessible and not visible. Phase two would then consist of the remedial works themselves, having had the benefit of Phase one to allow further investigation and properly specify the proposed works. If the works cannot be fully quantified and therefore the tendering contractors are unable to provide fixed prices, then another option is to award the contracts on the basis of agreed rates for carrying out various repairs. This could include a schedule of rates for different elements of work that can be quantified as the work proceeds, and a valuation for the amount of work undertaken agreed at the end of the contract. Whichever building contract and procurement route is adopted, the contract particulars should reflect the nature of the work, including all the aspects outlined in Figure 3.7.

Sometimes trials and tests of areas of repair will be undertaken to evidence that a particular repair method is working correctly before being applied over a much larger area.

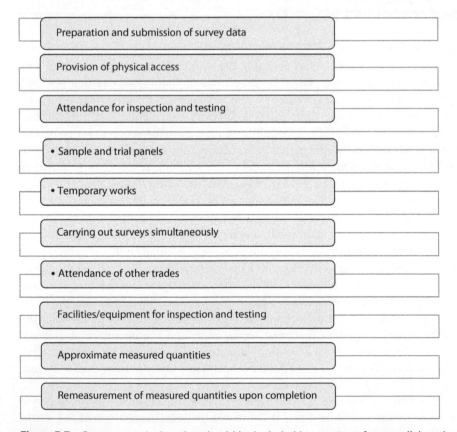

**Figure 3.7** Contract particulars that should be included in a contract for remedial works.

As an example, a façade repair contract could involve destructive removal of a small area of repair material to ascertain its adhesion to the underlying wall fabric. If the repair compound is relatively easy to remove by chipping off, then this might deem the repair system inadequate owing to a loss of key, and a new method may need to be devised. Other tests may be related to waterproofing where a small area of roof or walling is water pressure treated to ensure that the new material is fully waterproof before being deployed across a whole roof or wall area. In this way, it is far easier to test small areas and make adjustments where required to the design or specification of the repair system rather than assume a system will work and discover later there are widespread problems. Such trial areas are required sometimes to be undertaken at height or to areas that are difficult to access in other ways. In such cases, temporary scaffolding, cherry pickers or scissor lifts may be required to provide access.

During the remedial works it is essential to have levels of supervision to reflect the nature and complexity of the works. On specialist works a competent and specialist supervisor may have to be commissioned to undertake the inspections and to test the areas of repair where appropriate or deemed necessary. Faulty remedial works can sometimes result from poor workmanship or materials and good supervision should avoid the need for further remedial works. In addition, materials and repair systems should only be used where they are tried, tested and preferably accredited to a recognised national institute or body. In the UK, the British Standards Institute would be the relevant body for certification and accreditation of construction products and materials. In addition, contractors who are appointed to undertake such repair works should be selected on the basis that they possess the right level of expertise, with previous experience and a good track record of undertaking similar works. References should be obtained, with inspections of previous work that they have completed, if possible, but sometimes this can be challenging owing to time constraints in commencing the repair contracts. On completion of the remedial works, periodic inspections should be carried out to ensure that the repair system is performing adequately, and possibly to detect any early signs of failure that can be addressed before further failure emerges. In many scenarios the repair system should work in conjunction with building owners' maintenance systems to ensure that the required amount of maintenance is carried out as per the repair manufacturer's written instruction. In the case of a repair coatings to wall façades or roof covering this could involve specialist cleaning and/or periodic recoating at prescribed intervals.

## 3.7 Measures to Mitigate and Prevent Defects

Notwithstanding the array of information that is available for diagnosis and addressing building defects, there are still wide-ranging degrees of building defects that do frequently occur. In more severe cases such defects are left to develop unchecked without the necessary remedial works, which subsequently causes more damage, e.g., dampness and timber decay. In a similar way to the health sector, best practice in surveying should continue to be disseminated around prevention of defects as well as cure. Certainly, avoidance of defects are preferable to mitigating measures and strategies, and this can be achieved through regular building surveys as part of planned preventative maintenance policies and systems. The converse scenario would be reactionary maintenance whereby works are only planned when resultant damage from defects have been identified. An example could be water penetration through roof coverings, causing damage to ceilings and other finishes, which can incur more costs and increased disruption for building occupiers.

In addition to having adequate and effective maintenance management policies in place, defects can be reduced through good designs and specifications, which are geared to achieving longevity for buildings. Furthermore, ensuring quality control during the construction period for new buildings and refurbishment work can avoid latent defects being stored up for emergence at later stages. Guidance on maintenance planning and achieving performance of buildings through design and construction is widely available through professional bodies such as RICS alongside trade bodies, manufacturers and academic institutions. Such guidance is underpinned by a good knowledge of building pathology coupled with a comprehensive understanding of the science of the component, elements and materials which make up the structure and fabric of buildings. The science in this regard will be related to factors such as maintaining dimensional stability and a comprehension of the different ways that building components and materials move when subjected to varying degrees of temperature and moisture content. Other factors that the science will need to encompass include heat transfer mechanisms, air and water permeability through structure and fabric, relative humidity to avoid condensation forming and an awareness of the factors which cause corrosion of metal and timber decay. However, on an international scale, such guidance may vary from one country to the other depending on the predominant forms of construction in a particular country and the climatic conditions. For instance, what is appropriate construction for a desert climate, e.g., United Arab Emirates, may be very different from a building in an arctic climate such as Finland. Clearly, there will be different forces and climatic conditions imposed in different areas of the world with buildings in areas prone to earthquakes, flooding, extremes of temperature and heavy winds requiring special design and construction technologies to mitigate these factors. Consideration of such factors will affect the ways and means by which buildings are procured. In addition, it is important to consider at the design stages of buildings how they are going to be inspected and maintained post-completion. For instance, if a building has a high degree of glass curtain walling, it is imperative to consider how the glazing system is going to be regularly cleaned. In the same way where mechanical and electrical plant is located on roof areas, it is crucial to plan for a safe access system and procedure to be designed into the build that will allow qualified engineers to safely and easily access plant and from time to time to replace certain component parts. Where building elements are inaccessible, this can be problematic for inspecting their condition and for maintenance planning, as well as for arranging access for any remedial works. This can sometimes cause untold difficulties in faults emerging that escalate into more serious defects because of inspection restrictions. An example could be a roof space in an existing dwelling that has restricted or no access. As this confined and inaccessible loft space cannot be inspected there may be a long history of water penetration through the roof covering that has gone unnoticed and undetected for many years. This may have caused ancillary defects related to dampness in the timber structure of the roof, leading to widespread dry rot infestation throughout. The initial remedial works to rectify the roof problem may be insignificant compared to the vast amount of expense that dry rot eradication works and timber replacement may involve. It is a prerequisite of being a building designer, for all the above reasons, that they fully understand their client's future maintenance regimes and requirements at an early stage and allow for all the necessary provisions and equipment to facilitate their operations. In this regard they should consider worse-case scenarios where a fault or defect could arise that will require access for inspection and to carry out the remedial works.

In addition to the above, forward planning should take place, with consideration given to the location of elements which could be prone to failure in areas where they cannot be reached, or which could cause untold damage and disruption to a particularly sensitive and specialist area. An example of this could be not to site a cold-water tank or related pipework above a ceiling to a data communications room, as the consequential damage to the technology and the effects on business continuity in the event of leaking water could be immense. Furthermore, what may represent an initial additional capital cost in the build may actually save more across the life cycle of the building in not having to expend on temporary measures. One example could be that the installation of a gantry to access a roof atrium and thereby avoid the need for temporary scaffolding or the use of scissor lifts to access high areas of the space. In the UK, the Construction Design and Management (CDM) Regulations 2015 impose health and safety legal requirements on the whole design team and the clients to allow in the design for safe access of buildings with the focus on reducing accidents. On very tall buildings it is becoming more popular in recent years to install self-cleaning glass, which albeit more expensive than standard glazing, eliminates the need for manual cleaning, therein saving future maintenance costs and avoiding the potentially hazards of individuals working at height. The balance between initial capital costs and incorporating measures such as these that can reduce future maintenance costs will be discussed later in the book under life cycle analysis.

## 3.8 Summary

The pathology of building maintenance is related to the school of applied sciences and revolves around the investigation, inspection and reporting of building defects with principal aims and objectives around the need to comprehend the processes of building degradation. Such investigation work should identify the source or sources of building problems through diagnosis of the underlying technical issues and then determine the extent and specification of the required remedial works. In the UK, building pathologists are normally building surveyors, but if a particular building defect is related to major complicated technical problems, then specialist engineers may be appointed. When building surveyors are determining remedial works, following diagnosis of defects, it is important for them to focus on the root problem (cause) rather than simply address the symptoms of the defect. Certain building failures result from poor design and quality of build during the construction stage or refurbishment whereas others may simply arise due to natural forces associated with weathering and normal use. The condition of a building is not always correlated with its age, on the basis that some newer buildings can have more defects and signs of wear and tear than older buildings, as a result of poor design, workmanship and quality of build. Another factor relating to building condition and whether they are preserved or deteriorate, will depend on the degree of maintenance they receive over their respective life cycles.

The diagnosis of building defects is not always straightforward, as there may be more than one force at play in terms of multiple sources of different problems. In such cases, this can sometimes complicate normal lines of investigation and enquiry and present difficulties in determining the cause and associated remedial works. When considering the typology of defects in buildings one should distinguish between different types of problems, and these can be categorised by type and cause into (i) settlement and subsidence; (ii) condensation, dampness and water penetration; (iii) loss of bond and adhesion; (iv) decay and corrosion; and (v) movement and cracking. It is important to understand the nature and effect of agents that can lead to building defects, and these include electromagnetic, mechanical, chemical, building user, thermal and biological-related forces and effects. The most common defects arise from poorly manufactured materials, components and elements alongside poor workmanship in the construction process and it is therefore important wherever possible to only use materials and building components that are adequately accredited for the standards or quality. Another common cause of premature failure and defects is lack of general understanding and lack of maintenance of buildings. Accordingly, planned preventative maintenance systems should be instigated to address maintenance issues before they become more serious defects; there are still those who will only action works on a reactive basis. Furthermore, design team consultants should be diligent in their designs and specifications to ensure buildings are as maintenance free as possible and allow for access for inspections and maintenance works.

The repair or replace question frequently arises when considering remedial works and how building components and elements interact and work with each other. It should be considered whether a patch repair will be sufficient over time or merely be storing up a bigger problem for a later date. Sometimes remedial works do not solely look to repair or renew a particular element back to its former state, but to improve and adopt the design to make it function more effectively. Other considerations for the repair/renewal decision-making process could be grant funding for replacement works, disruption during the works, sustainability, life cycle costing and, clearly, affordability for building owners. Notwithstanding this premise, the options that surveyors and designers recommend should be arrived at through carefully considered mechanisms and consultation with their clients. Normally such recommendations would be accompanied by technical reports which clearly set out the different options, and the benefits, disadvantages and implications of each one. In a scenario where the repair or renewal costs are disproportionately high to the value of a building, clients may decide to demolish and redevelop, change the use of the building or simply sell the building and relocate elsewhere.

When considering remedial works to address defects and maintenance issues it is important to factor in the degree of complexity and wide-ranging specialist nature of the repair contract. For this reason, it is essential to only appoint appropriately selected main contractors, subcontractors and specialist contractors who are suitably qualified for the nature of work they are required to carry out. When the remedial works are agreed between the client and their supervising officer/surveyor overseeing the work, a brief outlining the works should then be prepared, with budget costings and a proposed plan of implementation with timelines. Remedial works will be closely managed by the supervising officer, normally the surveyor who is overseeing the works, and this will involve frequent inspections of the works and certification at various stages of the project. It may be necessary to have reviews of the work if there are unforeseen areas that cannot be adequately specified owing to access difficulties and/or opening up works via destructive intrusive investigation. Sometimes trials and tests of areas of repair will be undertaken to evidence that a particular repair method is working correctly before being applied over a much larger area. During the remedial works it is essential to have levels of supervision to reflect the nature and complexity of the works. On specialist works a competent and specialist supervisor may have to be commissioned to undertake the inspections and to test the areas of repair where appropriate or deemed necessary. On completion of the remedial works, periodic inspections should be carried out to ensure that a repair system is performing adequately and possibly to detect any early signs of failure that can be addressed before further problems emerge.

Notwithstanding the array of information that is available for diagnosis and addressing building defects, there are still wide-ranging degrees of building defects that do frequently occur. In some more severe cases such defects are left to develop unchecked without the necessary remedial works, which subsequently causes more damage, e.g., dampness and timber decay. In a similar way to the health sector, best practice in surveying should continue to be disseminated around prevention of defects as well as cure. Certainly, avoidance of defects through mitigating measures and strategies are normally preferable and this can be achieved through regular building surveys as part of planned preventative maintenance policies and systems.

# Reference

Thomas, M. (2013). *Managing Building Pathology and Maintenance*. Construction Managers' Library, Leonardo da Vinci: 2011-1-PL1-LEO05-19888.

# 4

# Maintenance Management and Performance Measurement as Part of Private Financing Initiative (PFI) Schemes

## 4.1 Introduction

While it is believed that there are other obstacles hindering the improvement of effective delivery of service in the healthcare sector, one essential element necessary to sustain progress in the field of facilities management is the ability to consistently and reliably monitor progress against the standard of delivery in all aspects including maintenance. Accordingly, performance monitoring, which encompasses the reporting and comparing of estates and facilities outcomes, including maintenance management, using key performance indicators (KPIs) is arguably one of the most important steps needed to deliver the quality of service expected in healthcare.

It can be argued that the delivery of healthcare maintenance and facilities services have organisational rigidities owing to the difficulty and effectiveness of the various monitoring and measurement parietals. For this reason, the varying standards of customers and stakeholders have made it difficult for maintenance and facilities services to create innovation or change. The question of value of service is both political and customer driven, and the larger the number of groups claiming to define values for facilities services in the healthcare sector, the more challenging performance measurement and making needed adjustments to the service become. This chapter will outline the dilemmas in this regard and offer background and context to this ongoing problem and present possible solutions.

The chapter starts with a background to the discussion on performance measurement (PM) of maintenance and other facilities management services in the healthcare sector. In this sense, it will articulate some of the problems that performance measurement initiatives have experienced in the past. Furthermore, it will also provide justification for highlighting the issue of performance measurement linked to PFI initiatives and explain the reasons why it is widely regarded as a 'hot topic'. Thereafter it will explore the definitions and concepts of facilities management, the application of performance measurement in a healthcare PFI setting and the impact of performance measurement on service and quality improvements. The significance, importance and challenges for PM around facilities management on a private financing initiative (PFI) scheme will then be discussed. This will include such issues as the financial dilemmas facing the healthcare sector in the UK and how it has been tasked to achieve more with less, and targets to improve the quality of the health service while reducing cost. Other challenges for maintenance and facilities management will be covered, including the need to increase accountability, drive up quality, meet increasing demands on maintenance and facilities outcomes, make efficient savings and re-energise value-for-money approaches. In the context of such challenges further explanation will be presented on why these factors have all led to renewed interest in the direction of performance measurement. In addition, it will also explore payment mechanisms as one of the underlying reasons for measuring performance in a PFI healthcare project as a gain share/pain share initiative. In this sense it will explain the issues around deduction of payments for poor performance and, conversely, financial incentives for exceeding targets. Various performance measurements for improving maintenance and facility management outcomes tools are then explored alongside the role of KPIs. The benefits of performance measurement will be articulated from a position of presenting a clear picture of where improvements are actually happening.

Justification for introducing performance measures on PFI schemes and the need to assess progress against predetermined objectives will be covered. This will include measures to identify areas of strength and weakness and the need to align future initiatives with the aim of helping to improve the organisational performance. The chapter will then extend into performance measurement as applied to quality improvements in maintenance and facilities management practices. Financial and non-financial measurements, being the primary variables that can be used to research the construct of performance measurement in most organisations, will then be examined. The various forms of performance monitoring tools will be identified and analysed, and shown how these relate to benchmarking and KPIs. Finally, key issues arising for performance management as part of a Facilities Management (FM) tool on PFI schemes will be articulated and discussed, alongside reflections for the future.

## 4.2 Definitions and Concepts of Facilities Management

Facilities form important elements in the delivery of a successful healthcare service and represent a large part of the organisation's assets and expenses (Barrett and Baldry, 2007). As stated earlier, the provision of an FM service in a healthcare organisation has an impact on the overall quality and effectiveness of the services it delivers (Shohet and Lavy, 2004). Furthermore, the British Institute of Facilities Management (BIFM) defines FM as 'the integration of multi-disciplinary activities within the built environment and the management of their impact upon people and the workplace' (British Institute of Facilities Management 2003), while the International Facility Management Association (IFMA) defines FM as 'the practice of coordinating the physical workplace with the people and work of the organisation' (International Facility Management Association, 2003).

FM is also defined as 'an integrated approach to maintaining, improving and adapting the buildings of an organisation in order to create an environment that strongly supports the primary objectives of that organisation' (Barret 2000). 'The application of integrated techniques to improve the performance and cost effectiveness of facilities to support organisational development' was the definition provided by Shohet and Lavy, (2004). Furthermore, Barrett and Baldry, (2007) described facilities management as 'an integrated approach to maintaining, improving and adapting the building of an organisation in order to create an environment that strongly supports the primary objectives of that organisation'.

## 4.3   Background to the Discussion on Performance Measurement (PM) of Facilities Management and Maintenance in the Healthcare Sector

### 4.3.1   Definitions and Concepts of Performance Measurement

Measurements are needed by an organisation to assess performance against predetermined objectives, and this is particularly useful for meeting maintenance management and facilities management objectives and targets. It provides information necessary for management to make intelligent decisions on the objective of the service (Amaratunga and Baldry 2002a). A performance measurement system reflects, amongst other things, the needs of the customers as well as the organisation's; a set of rules for decision making; an understanding of the process and the application; the consistency of the system; the interpretation and compatibility with the existing systems; the precision of use, methods of interpretation data: and the economic importance of the process (United States Department of Energy (TRADE) 1995).

According to Hatry (2006), regular measurement of progress toward specified outcomes is a vital component of any effort at 'managing-for-results'. The main function of any performance measurement process is to provide regular, valid data on indicators of performance outcomes. However, as argued by TRADE (1995), performance measurement is not simply concerned with collecting data associated with predefined goals and objectives; rather, it is considered as an overall management system which involves prevention and detection for the purpose of delivering the required service to the organisation and its customers. Consequently, Armstrong and Baron (2009) argued that PM is the outcome of an activity that determines the progress of an endeavour by helping to analyse how well the service is performing, if it is meeting the set objectives, if the customers are satisfied, if the processes is appropriate and where improvements may be necessary. In addition, performance measurement also analyses the success of a work group, programme, or organisation's effort by comparing data on what actually happened to what was planned or intended (Pratt et al. 1997). In this sense, performance measurement is the 'process of quantifying the efficiency and effectiveness of an action' (Amaratunga and Baldry 2002b).

Perrin et al. (1999) stated that 'performance measurement is the selection and use of quantitative measures of capacities, processes, and outcomes to develop information about critical aspects of activities, including their effect on the public'. Performance measurement asks 'is progress being made toward desired goals? Are appropriate activities being undertaken to promote achieving those goals? Are there problematic areas that need attention? Are there successful efforts that can serve as a model for others?' (Perrin et al. 1999). Seemingly, the measurement of performance seeks to monitor, evaluate and communicate the point at which individual aspects of the service conforms to the key objectives of the healthcare organisation (Hatry 2006).

Neely (1999) maintained that the main reasons for the development and implementation of performance measurement for facilities and maintenance management services, including changes in organisational roles and stakeholders' demands, are increases in competition and the power of information technology, which has created more informed consumers. Nani et al. (1990) stated that performance measurement systems came about as a means of monitoring and maintaining the organisational processes, allowing an organisation to pursue the achievement of its core business objectives. Spitzer (2007) stated that PM came about because of the quest of human needs for social interaction, the ambition to amalgamate the world better in terms of trade and commerce.

In their study, Amaratunga and Baldry (2002b) stated that the development of performance measurement was influenced by the general push for better quality of service while meeting other cost variables. The authors opined that measurement is only a 'means' and not an 'end' because results of performance measurement provide the basis for an organisation to assess how well it is progressing towards a predetermined goal, identifying areas of strengths and weaknesses, and deciding on future initiatives, with the aim of improving organisational performance. Bititcti et al. (2000) identify that performance measurement processes need to embrace changes in both the internal and external environment of the organisation by 'reviewing and reprioritising internal objectives when the changes in the external and internal environment are significant enough; deploying changes to internal objectives and priorities to critical parts of the organization, thus ensuring alignment at all times; and ensuring that gains achieved through improvement programs are maintained'.

### 4.3.2 Evolution of Performance Measurement (PM) in Healthcare Facilities Management

Performance measurement (PM) in healthcare facilities management (FM) services has evolved since the days of Florence Nightingale who, in the nineteenth century, was concerned about sanitary conditions in military hospitals during the Crimean War and in London. She developed a method of data collection and a statistical analysis system focusing on in-hospital mortality, which allowed her to compare units within hospitals over a period of time. The measurements system allowed Florence Nightingale and others to make significant breakthroughs in exploring the relationships between sanitary conditions and the mortality rate (Nerenz and Neil 2001).

What makes a hospital's facilities management special is the need for the provision of 24/7/365 days round- the-clock facility services. Failures in the provision of FM services in a hospital could have dramatic consequences, and is a characteristic which represents unique operating conditions involving much greater stakes than any other type of establishment (Lennerts et al. 2003; Ventovuori 2006). Davies and Lampel (1998) highlighted that a major challenge facing the healthcare system is controlling costs while improving quality and increasing access. Consequently, if the success factor is to improve quality as well as control cost, then an effective measure of quality is needed to inform the customers. According to the Health Foundation (2013), quality of care is difficult to measure and complex to define. The 'rigorous and consistent measurement and assessment is therefore challenging; and there are significant limitations to the data available, as adequate indicators do not exist for each domain of quality and each service'. Loeb (2004) identified one of the challenges of measuring performance in healthcare as being the unyielding attitude and varied notion amongst key staff who believe that performance measurement is too fraught with problems to be useful. Some would regard performance measurement as a costly endeavour that, in the absence of the availability of electronic data capture, does not provide sufficient cost incentive. Measurements are also portrayed as expensive, time consuming and capable of provoking anguish, worry, anxiety and frustration amongst the people being measured, and the people doing the measurement. The complex nature of the healthcare system, beleaguered with a variety of salient issues and challenges, makes it difficult to measure the effectiveness of the quality of service on offer (Loeb 2004).

Purbey et al. (2007) postulated that performance measurement in other business organisations is nothing compared to the increasing levels of competition in the health service, joint ventures, patient services and quality initiatives, emphasising continuous quality improvement as a cornerstone of the operations of a healthcare organisation. The advent of digital technology and the increasing knowledge of consumers have further intensified demands to have information available in order to aid the appropriate healthcare decisions. The healthcare system is composed of a set of complex entities, activities and processes at the core of which are the inevitably clinical processes involving a wide range of participants, bringing to the system a different set of needs, priorities and evaluation criteria (Gopal and Patricia 2003). Accordingly, the objective of the research undertaken by Lyere (2013) was to investigate the role and effectiveness of the performance monitoring mechanisms around facilities management and to identify the factors affecting service delivery in a PFI hospital. Various means of performance measurement tools were considered but the role of KPIs was reviewed in the context of this research to know if the application of KPIs as a measurement tool:

- Leads to better performance.
- Drives quality and innovation.
- Is effective and appropriate in application.
- Is a valuable quality assurance tool or just part of a compliance target-driven culture. Demonstrate if it provides an effective form of incentive for the service provider.

### 4.3.3 Justification for the Research into PM Applied to a Hospital PFI Contract

We know, for instance, that we have to measure results. We also know that with the exception of business, we do not know how to measure results in most organisations.

(Peter Drucker, *The Age of Discontinuity*, 1968).

The fundamental objective of any healthcare organisation is to achieve good outcomes for the patient. Achieving good progress in an estates and facilities maintenance context requires strong measurement capabilities to guide quality improvement initiatives, support decisions on what work best and foster the development of continuous learning process. Performance monitoring is a required element in the healthcare FM service because it provides an excellent and efficient mechanism for improving productivity (Mecca [1998] cited in US Department of Health and Human Services [2006]). According to Berg et al. (2002), the PFI agreement can only work if the outputs and inputs specified in a contract can be effectively measured, making it a grey area for public services organisations that are multidimensional with a quality that is difficult to assess.

Effective management is based on the foundation of an effective performance measurement. Measurement can help the client management to identify whether or not the service provider is meeting the requirements of the customers. It ensures that decisions are based on facts, not on intuition, speculation, perception or emotion. Performance measurement in FM services is necessary in order to assess whether progress is being made towards desired goals and whether appropriate programmes of activities are being undertaken to promote the achievement of the objectives (Amaratunga and Baldry 2002b). Managing service performance in a PFI hospital is also fundamental to the contract, besides the fact that it provides a vital support for the core operations, it is also the process by which payments to the service provider are calculated and any necessary deductions made. According to Partnership UK (2006), managing a PFI service performance involves monitoring the achievement of the service outputs to ensure that the contractual performance and improvement processes are maintained. The success of a PFI project is also dependent on effective performance monitoring because it provides incentives for improvement and delivery of service to specification even though there may be complexities like inadequate resources for performance monitoring, misinterpretation of the output specification and the level of deductions compared to the actual services delivered (Herbert and Jon 2009).

Healthcare is a vital service that touches people's lives daily by providing treatment and resolving the health problems of the patients through the staff. Ultimately, human lives are dependent on the skilled hands of the staff and those who manage the infrastructure that supports the daily operations of the service, thus making it a compelling case for research (The Department of Justice 2004). Furthermore, the challenges facing all health trusts in the face of the economic downturn, current global competition and increasing demands from stakeholders, is a distinct need to improve healthcare service performance and deliver excellent service. Invariably, this has thus led to various methods of performance measurements emerging as quintessential tools to optimise the pursuit for accountability in the delivery of FM services (King's Fund 2010). Accordingly, improving the quality of health services is now a key requirement within the National Health Service, supported by key initiatives such as the following:

- Commission for Quality and Innovation (CQIN)
- Care Quality Commission (CQC)
- Patient-Led Assessments of the Care Environment (PLACE)
- National Health Service quality standards
- Patient Experience Tracker Dashboard
- Matron's Charter
- National Standard of Cleanliness

### 4.3.4 The Significance, Importance and Challenges for PM around Facilities Management on a PFI Scheme

The healthcare service in the UK via the National Health Service offers services to a diverse population that is complex in nature, diverse in operation and distinctive in services which are provided mostly free of charge. There is the diversity of income, education and employment; culture, language and ethnicity; age, physical and mental conditions; and hopes, dreams and goals. Providing services to all of these people require that patient service providers need to understand the social and environmental circumstances of their lives; demographic origin, cultural beliefs and values regarding health, health needs and effective ways to communicate with them (Lichiello and Turnock 1999).

Strong measurement capabilities are the requirement to achieving good progress in healthcare services, which could support decisions on what works best, guide improvement efforts and promote the development of a learning objective. In spite of the problems associated with the improvement of the healthcare system, the ability to consistently and reliably measure progress across all aspects of patients' health, both clinically and non-clinically, remains an important factor for sustained progress (Porter 2008). The financial dilemma facing the healthcare service is how to achieve more with less; how to improve the quality of the health service while faced with the challenges of reducing cost (The Economist 2013). The need to increase accountability, drive up quality, meet increasing demand posed by the ageing population, respond to rising patient expectations, make efficient savings and re-energise a value-for-money approach, have all led to renewed interest in the direction of performance measurement (Wilcock and Thomson 2000; The Health Foundation 2009).

Different healthcare trusts are now struggling to find ways of using performance measurement to promote improvement in the delivery of care for patients and carers. Health trusts are being asked to account for their facilities management performance and various methods of performance measurement have therefore emerged as quintessential tools to optimise the quest for accountability (Martinez 2000). According to Neath (2010), effective measurement systems and tools are needed to help the health service meet the burgeoning challenges of improving performance and reducing cost. Neath acknowledged the need for the NHS to apply performance measurement to all its activities to demonstrate quality improvements and to determine whether efficiency savings have been made in the service delivery. If the service is to compete favourably, it needs to ensure that there is a correlation between the cost of the service provided and the desired improvements in patient and staff experience, satisfaction, cost effectiveness, safety, outcomes, quality of care, prevention, health population and staff productivity.

Øvretveit (2003) highlighted that poor quality of facilities services, including maintenance, wastes money that could be used to treat patients or fund research programmes, making it a compelling case for stakeholders' clamour for demonstrable evidence for quality improvement. In view of the occurrences, healthcare trusts are expected to proactively introduce and monitor quality process and other strategies to ensure patients safety and improved quality of service. Neath (2010) provided that the health service needs to collect quantitative and qualitative data over an extended period and at regular intervals to form decisions and to assess whether resources are meeting the desired goals.

Highlighting examples of excellence in practice of where performance measurement systems have been aptly applied and have a real impact on improving patient care, Neath opined that staff attitude to performance measurement was a major challenge facing the organisation.

The primary function of the National Health Service (NHS) is the ability to provide an acceptable quality of care to the community, the staff and the stakeholders it serves (Heavisides and Price 2001). The need for capital investment and improvement in the delivery of these services, which encompasses maintenance management, has given rise to a need for the involvement of private sector partners in the provision of modern facilities (Holmes et al. 2006). According to Her Majesty's Treasury (HM 2003), the use of PFI for the procurement of NHS projects is to bring together the added advantages of the private sector's management skills, commercial expertise and the discipline to the delivery of public infrastructure. Under a PFI contract, the public sector client enters into an agreement with private sector companies through a consortium to design, build, finance and operate the asset of the hospital for a fixed period (NAO 2005). It is claimed that the use of PFI in the NHS ensures improved delivery of projects with respect to time, cost, quality and improved maintenance of public infrastructure (Dixon et al. 2005). Underpinning the PFI project is the provision of both the hard and soft services under the term 'facilities', which is guided by a service-level agreement (SLA) to regulate the operational performance of the service contracts. According to Shohet and Lavy (2004), the provision of an FM service in a healthcare organisation has an impact on the overall quality and effectiveness of the services it delivers. Supporting this view, Gelnay (2002) claimed that a healthcare FM service is a key element in the delivery of a successful healthcare service. Healthcare facilities management is considered one of the essential elements in the delivery of successful healthcare services because failures in the provision of FM services in a hospital could result in far more dramatic negative consequences than in any other general type of building (Gelnay 2002; Ventovuori 2006).

Partnership UK (2006) proclaimed that managing a PFI service performance involves monitoring the achievement of the service output to ensure that the contractual performance and improvement processes are maintained. Managing service performance in a PFI hospital is also fundamental to the contract; besides the fact that it provides a means by which payments to the service provider are calculated and necessary deductions made, it also provides a vital support for the core operations of the organisation. Achieving value for money in a PFI project is largely dependent on performance monitoring, as it ensures that service delivery is in accordance with the output specification (Robinson and Scott 2009). To some extent, performance monitoring provides an incentive for the contract management to deliver the standard of services required by the client management as stipulated in the output specification (Ng and Wong 2007).

Seemingly, Leahy (2005) opined that a rounded view of overall performance is needed from the procurement process to the operational performance of the facilities of a PFI project. In agreement, McDowall (2000) stated that it is crucial to have an effective performance mechanism to assess compliance with the service level agreements due to the fact that PFI projects are based on a premise of effective and efficient delivery of services to the customers. The National Audit Office publication (NAO 2005) on the performance and management of hospital PFI contracts, emphasised that an appropriate means of measuring performance that is directly linked to the payment regime is crucial in PFI projects as it ensures that operational risks are tested, and the financial penalties applied for service failures. Performance measurement in FM services is therefore necessary in order to assess whether progress is being made towards the desired goals and whether appropriate programmes of activities are being undertaken to promote the achievement of objectives (Amaratunga and Baldry 2002b). In addition, performance measurement can also assist in the identification of service areas needing attention (Perrin et al. 1999). Kennerley and Neely (2003) observed that the events over many years have necessitated new challenges in FM to improve the quality of service, with many organisations redesigning their performance measurement systems to embrace the status quo.

### 4.3.5 The Benefits of Performance Measurement

TRADE (1995) identified that the measurement of performance in a maintenance management context helps organisations determine if they are achieving the customers' requirements; provide an understanding of the process; confirm what the management is aware of, or reveal what they unaware of; and identify where improvements are needed. It ensures that decisions are based on facts and figures, not on emotion or perception. It helps reveal the emotional bias, longevity and cover-up associated with the process. Finally, a successful performance measurement provides a clear picture of where improvements are actually happening. If staff have been doing their job for a long time without measurements, they may assume that things are going well. 'They may or may not be, but without measurements there is no way to tell' (TRADE 1995).

### 4.3.6 Justification for Introducing Performance Measures for Facilities Services and Maintenance Management on PFI Schemes

Performance measurement (PM) is an important component for success, and processes established on measurable objectives are likely to be more efficient and effective than those that are not (McIvor et al. 2009). Without dependable measurements, organisations may not be able to quantify progress, monitor outcome and adjust process to produce desired objectives (Neely et al. 1995). The authors described performance measurement as 'the process of quantifying the efficiency and effectiveness of action and a metric used to quantify the efficiency and/or effectiveness of an action'. The management guru Peter Drucker is often quoted as saying that 'you can't manage something if you can't measure it' (Drucker 1998). Drucker, hailed by *BusinessWeek* as 'the man who invented management', believed that organisations have no way of knowing if they are successful unless their objectives are defined and tracked (The Drucker Institute). Quantifiable measurements are needed in facilities management because they provide proof of progress and success and can also identify areas needing improvement.

According to Welch and Mann (2001), PM is one of the cornerstones of business excellence which encourages the use of performance measures, more specifically, in the design, to ensure that the methods are aligned to strategy, and that the system is effective in communicating, monitoring and driving performance. Lichiello and Turnock (1999) summated that 'in order to improve something, you have to be able to change it; in order to change it, you have to be able to understand it and in order to understand it you have to be able to measure it'. Amaratunga and Baldry (2003) described performance measurement as the process of assessing progress towards achieving predetermined goals, including information on the efficiency by which resources are transferred into goods and services. Lichiello and Turnock further argued that performance measurement should not be used as a vindictive exercise, but as something to be done in partnership in order to achieve common goals. Measurements provide information for an organisation to assess progress against predetermined objectives, identifying areas of strengths and weaknesses and aligning future initiatives with the aim of helping to improve the organisational performance (Amaratunga and Baldry 2002a). They can also provide a deeper understanding of performance. Quality expert James Harrington (cited in Spitzer 2007, p. 19) said 'if you can't measure something, you can't understand it. If you can't understand it, you can't control it. If you can't control it, you can't improve it. In any area of management, without good measurement, it is impossible to know what is working and what is not working'.

Kincaid (1994) summated that society has become obsessed with measurement and the measurer of everything, making it one of the most prominent features of human endeavour, ranging from hospital, business, politics, economics, education, etc. Society ranks hospitals, universities, colleges, sports leagues, financial worth and almost anything that can be measured, and universities and hospitals pride themselves about being in the top performance league. An effective performance measurement system provides a platform from which client management can assess the progress of the service to the agreed specifications in a PFI contract; help identify areas of strengths and weaknesses and enable well-informed decisions to be made as to how to improve performance (Amaratunga and Baldry 2002a). Performance measurement 'seeks to monitor, evaluate and communicate the extent to which various aspects of the health services meet their key objectives' (Hatry 2006).

Performance measurement informs quantitatively something important about product, services, and the processes that produce them. They are tools that help the organisation understand, manage, and improve the processes of the organisation. TRADE (1995) highlighted that performance measurement provides information on:

- How well the organisation is doing
- If it is achieving its goals
- If the customers are satisfied
- If the processes are in statistical control
- If and where improvements are necessary.

The data collected provide the necessary information for the organisation to make intelligent decisions about the service it offers. A performance measurement is the combination of a number and a unit; the number gives the magnitude (how much) and the unit gives the number a meaning (what). Most performance measurements are expressed in units of measure to make meaning to those who use them for their management decisions (TRADE 1995). Facilities management therefore requires both quantitative and qualitative measures of performance to enable it to compete favourably.

## 4.4   The Advantages and Disadvantages of PFI Ventures in a Facilities Management Context

Berg et al. (2002) argued that one of the major advantages of a PFI venture is the allowance provided for the fiscal deficit implications of large infrastructure projects to be smoothed over time, despite the fact that it is associated with a transfer cost to future government. Equally, a major disadvantage of PFI contracts is their relative inflexibility, making it an expensive venture to break away from if it proven not to be meeting societal needs. Also, both parties face potential hold-up problems if one of the parties decides to to increase the cost to the other party by taking advantage of changing circumstances that are not specified in the original contract (Berg et al. 2002).

Lyere (2013) reported that in 2012 there were 717 current PFI projects in the UK, of which 648 were operational. The total capital costs of current PFI projects at that time was £54.7 billion (HM Treasury 2012). A BBC Southwest Home Affairs programme 'Private Finance Initiative: PFI projects cost £2.4bn' quoted the economist Kevin Butler, a former adviser to the Bank of England, that it was not surprising that the PFI cost more. He stated that 'It's always cheaper for the government to borrow itself, rather than to use the private sector. In terms of borrowing costs, it also makes more sense for government to borrow to fund these programmes, rather than rely on private finance, especially at a time when private finance is in short supply'.

## 4.5   Quality Improvements in Maintenance Management Brought about through Performance Measurement

Quality is described as a particular feature of a good or service in terms of its hardness and colour which cannot be evaluated by itself, rather, it is observed in terms of the expectation the desired good or service is expected to fulfil ... 'care is effective, when it is able to produce the expected effect; it is efficient when it is produced faultlessly, free of malpractice; and it is optimal, when the provided solution corresponds to the state of the art' (Nies et al. 2010). Quality is multidimensional and can be difficult to measure, define and observe because of the individualised nature. Quality is better viewed or judged from the 'eye of the beholder' (Nies et al. 2010). A famous statement attributed to Florence Nightingale appropriately captures the relationship between performance measure and quality management: 'the ultimate goal is to manage quality. But you cannot manage it until you have a way to measure it, and you cannot measure it until you can monitor it' (Arah et al. 2003).

The World Health Organization (WHO 2006), in the context of healthcare, identified six dimensions of quality: effective, efficient, accessible, acceptable/patient-centred, equitable and safe. Common themes such as effectiveness, efficiency, safety and experience of care of recipients, exists within the different definitions of quality. For example, the Organisation for Economic Cooperation and Development (OECD 2009) also focused on effectiveness, safety and patient-centredness as the rudiments of efficient healthcare system. The Department of Health (DOH 2010) also defines quality as a composite of four factors: effectiveness – achieving the best outcomes and getting it right first time; efficiency – delivering value-for-money services; experience – service users receiving a positive experience of care and support from the service provided; and safety – protecting vulnerable people from any health-related harm.

There is a fundamental difference between standards of services delivered, the outcomes that people expect, and the satisfaction received in relation to their own well-being. It is a common fact that the customer is the best judge of quality and should therefore relate to what is delivered, rather than relate to the result of the delivery in terms of the effect on the user's sense of well-being (Goodrich and Cornwell 2008). The fact that a service provider is providing a service to an agreed and measurable standard does not necessarily mean that the customer is getting the best out of it, or in a way that they would expect the service to fulfil their needs (Goodrich and Cornwell 2008).

The book *In Search of Excellence* by Peters and Waterman (1982) emphasised on the qualitative aspects of a businesspeople, customer satisfaction, 'nurturing of unruly champions and managing by wandering around'. Peters and Waterman believed that in the customer arena, regular measurement of customers' satisfaction provides a better way forward for future organisational health than other market factors. To enhance competitiveness, Peters and Waterman suggested that companies should join in partnership, rather than in adversarial relationships, with suppliers and customers. A number of high-profile cases have been highlighted, including the Care Quality Commission (CQC) report, into the failings of some NHS hospitals in England (CQC 2013). According to Ellis et al. (2010), unlike the private sector, there is little use of standardised FM practice specifications within the NHS and the traditional outsourcing method often makes no provision for effective performance measurement standards. Failure in healthcare service has a huge impact on people's lives, and quality remains a serious concern in the NHS with expected outcomes not predictably achieved and issues concerning wide variations in standards of service delivery (King's Fund 2013).

Without effective measurements that integrates all the elements of quality, the service users have no way of knowing 'what lies beneath' the care they are given. In the words of Spitzer (2007) 'the business imperative today is not just to perform excellently but to perform excellently consistently; in the absence of good measurement, it is human nature to pay attention to the unusual or the annoying. That is why squeaky wheels often get the grease, even if it is the wrong wheel'. Longenecker and Fink (2001) advised that organisations that do not incorporate continuous performance improvement into their management structures, risk poor performance of service, customer dissatisfaction and high employee turnover. Besides everything else, the fundamental objective of monitoring an FM service in a healthcare setting is to improve the quality of service to the patients and also provide a basis for continuous improvement process in line with the specifications and the monitoring process. Therefore, every initiative taken to improve the delivery of FM services in a hospital should have some understanding of 'quality' attached to it. It is not desirable to design an intervention and measure to improve a process without an understanding of the quality element that drives the service in the first instance (Spitzer 2007).

Quality, as described by Edward Deming, is the degree to which performance meets expectations (cited in Chandrupatla 2009). According to Deming, 'Good quality means a predictable degree of uniformity and dependability with a quality standard suited to the customer' (Chandrupatla 2009). The American Society for Quality (ASQ) adopted the definition that 'quality denotes an excellence in goods and services, especially to the degree they conform to requirements and satisfy customers'.

The Institute of Medicine (2001) defines quality as 'the degree to which health care services for individuals and populations increase the probability of desired health outcomes and are consistent with current professional knowledge of best practice'. Quality in healthcare means providing the right services, in the right way, every time for every patient. The ultimate aim of designing a service is to ensure that the end user is satisfied with it. Therefore, the test and the evaluation process should lie with the consumer whose needs must be taken into consideration and translated into measurable characteristics (Spitzer 2007). The Oxford Dictionary meaning of quality is 'the degree of excellence of something'. This view was further highlighted by Professor Sir Mike Richards CBE, Chief Inspector of Hospitals, on a Quality Watch conference that the CQC inspects, monitors and regulates various services in the healthcare to ensure that the services meet the quality expected by the customer (CQC 2013). In essence, quality which is a conformance of standards and requirements should be measured by the degree of satisfaction felt by the customer.

## 4.6 Financial and Non-financial Measurements

Financial and non-financial measurements are the primary variables that can be used to research the construct of performance measurement in an organisation. Pavlov and Bourne (2011) relayed that the practice of management accounting in the second half of the twentieth century paved the way for the use of performance measurement to monitor performances in the organisation. In his opinion, Otley (2002) provided that the last two decades saw a shift from purely accounting measures with a great deal of interest now being paid to the development and use of non-financial measures which report on service performance as well as stimulates.

Financial measures rely on the organisation's financial information recorded in the income statement balance sheet and statement of cash flow; however, growth measures and non-financial measures, popularly referred to as operational measures. are the variables that represent how the organisation performs in non-financial issues, e.g., stakeholder perception and customer satisfaction. Most of the measures in this category rely on monitoring data from service providers as an assessment of their performance and may sometimes be subjected to different interpretations (Pavlov and Bourne 2011). There are advantages and disadvantages to the use of financial and non-financial measures when assessing the performance of an organisational process. It is incumbent on this researcher to select the operational measure which is the non-accounting measure that captures the essence of the organisational performance, given the environmental circumstance of this case study.

## 4.7 Performance Management

Performance Management is mainly concerned with managing the expectations of the organisation. Equally, it also determines how the achievement of the expectation will be measured (Armstrong and Baron 2009). Hurst and Jee-Hughes (2001) described PM as 'the whole set of institutional and incentive arrangements by which performance information is (or is not) used to influence performance in health care systems'. However, OECD (2009) argued that *'creating a PM system does not in itself improve performance. Its success, in part, depends on goals and strategies being clearly defined and communicated to employees, and on managers' ability to objectively assess and measure performance'.*

As noted by Hatry (2006), performance measurement is a recurring process; for it to be meaningful the organisation needs to benchmark and compare its activities with the benchmark of another organisation over time. Unlike performance measurement that is only concerned with outcome measurement, PM is able to respond to changing needs of the stakeholders as their taste and demand changes. In order for an FM organisation to make effective use of performance measurement results, it must be able to migrate from measurement to management, anticipating the necessary changes in line with the strategic goal of the organisation and a modality to effect strategic change. The accomplishment of these response tasks represents the foundation of good PM (Amaratunga and Baldry 2002a).

Performance measurement focuses on measuring outcomes and efficiencies from data obtained from benchmarking, including the use of key performance indicators (KPIs). It is essential to record such data in a form that easily facilitates audit and tracking of performance outcomes within an organisation. According to (Hatry 2006), if the data is not robust or collected in a form that does not allow comprehensive analysis, then the effort and cost of the performance measurement process will be wasted.

## 4.8 The Challenges for Performance Measurement

Mixed research exists as to whether measurement of performance has a positive or negative impact on the delivery of FM services. While some research favours either or both, others have demonstrated that there are no benefits derived from using performance measurements to judge organisational processes. For instance, Berwick et al. (2008) pointed out that performance measurement has a tendency to focus only on a particular service, which is a narrow aspect of the complete service. The tendency is for it to encourage improvement only on those areas being measured as opposed to the broader objectives of the complete process; therefore, the quality of service that is delivered, the engagement of patients, and the service received may be compromised. Services in a healthcare setting are intertwined; changes to any of the areas may have a knock-on effect on the others. Furthermore, there are other factors that can influence a service user's perception of a particular service, many of which may lie outside one particular service (McGinnis et al. 2002). Furthermore, Porter (2008) stated that performance measurement which is an outcome measurement tends to focus on the immediate results of a service or intervention, as opposed to the overall success of the full care cycle of the patient. Outcomes are the results of a service in terms of the care of the patients over time. 'They are distinct from care processes designed to achieve the results, and from key indicators that are predictors of results' (Porter 2008). Performance measurement should be a customer-focused concept that looks at maximising benefits and minimising negative consequences for the end user. Newcomer (1997) summated that performance measurements focus on measuring what is occurring but does not ask 'why' or 'how' it is occurring.

Consequently, Amaratunga and Baldry (2002b) summated that measurement should provide the basis for an organisation to assess how well it is progressing towards a predetermined objective, given that it is not an end, but a tool to aid effective management which may not necessarily indicate what happened or why it happened, or what to do about it. Expressing their views on the pitfalls in performance indicators, Davies and Lampel (1998) stated that the approach to PM only detects errors at the end of a process, rather than built in quality, which can often result in delays in acting on the data findings. In such a practice, the patient discovers that their problems are only corrected after their discharge from the hospital.

Neath (2010) ascertained that communication is one barrier to the effectiveness of PM in the NHS because of the way it is presented; staff cannot relate the benefits to the patients or visualise the importance to their work. Neath (2010) added that measurements have to be meaningful rather than being regarded simply as data against which staff can be judged. To encourage the use of PM and build it in to all daily activities, staff need to establish its relevance to their own work including the benefits of measuring, especially in terms of how it relates to improving the care experience of their patients. There is also the notion of fear and judgement associated with performance measurement. Spitzer (2007) opined that those negative pressures sometimes propel people to do whatever it takes in order to comply with the expectations of performance measurement. Explained Edward Deming, even if it means attaining a particular score to avoid failure, people will do whatever it takes to meet targets 'even if they have to destroy the organisation to do so'.

As cited in Spitzer (2007, p. 27), 'never underestimate how clever frightened human beings can be when faced with numeric targets'. There is also the misconception that performance measurement is synonymous with judgement, and staff may feel that their performance is being measured so that they can be judged, which negates the need to innovate and do things differently by taking risks and accepting that things may not always go well the first time (Neath 2010).

Servicer users' perception of measurement can be divided into two broad categories: the clinical and non-clinical. While clinicians share concerns about how measurement might affect their clinical practices as it focuses on important clinical processes, non-clinical service on the other hand, are typically concerned with the impact and value of measures on the services provided, favouring composite measures that focus on outcomes for allied financial reasons, rather than measures of process, compliance, and the well-being of the patients (Porter 2008). Taking a holistic perspective of both the clinical and non-clinical services, a combined hierarchy of measures can be developed and linked to evidence-based processes, followed by outcome measures supported by acceptable processes measurements that have a proven impact on the outcomes of the service provided. The development of these preferred measures requires substantial evidence, and, at present, there is a dearth of high-quality, consistent data to guide the implementation and validation of these measures (Porter 2008).

It was further clarified by Porter (2008) that in order to drive quality and innovations in healthcare services, 'outcomes should be measured continuously for every patient, not just retrospectively in the context of discrete approach or evaluations. Whenever possible, outcomes should be measured in the line of care and inform continuous learning. The current approach to outcome measurement is skewed toward retrospective studies, usually focused on a single end point. The bias towards these methods is one of the reasons that outcome measurement remains so limited, despite its overwhelming benefits'.

The delivery of FM services from the customer-related process was assessed by Amaratunga and Baldry (2002b) and they contended that the service needs to have a retrospective view of the customer targets and align outcomes to meet the desired targets. The authors provided three success criteria to ensure customers' expectations are continually met: the degree of communication and partnership from the service team; quality of the service delivered; and timeliness of the service delivery. A different outlook on the customer process from the perspective of the staff working in the organisation was provided by Simons (2000) who claimed that the staff of the organisation knows the service best, the customers best, and how best to improve the service. They know the difference between right and wrong and want to contribute to the success of the service, which they can be proud of. Therefore, PM and the monitoring mechanisms cannot be designed without taking into consideration the human element, and the cause and effect of the organisational processes. The misuse of measurement can trigger a bad relationship because of its enormous power. It is like a focal lens through which performance is viewed, triggering the wrong action if focused to the wrong direction; without a sense of vision for the positive and the negative sides of performance measurement, one might be inclined to have a false sense of satisfaction believing that any problem can be detected by routine monitoring (Simons 2000).

Theriault (2010) opined that performance measurement should not only be all about KPIs to track results and punish failures; instead, results of KPIs should be used to effectively manage the service. O'Leary (1995) asserted that measurement can be a loaded gun – 'dangerous if misused and at least threatening if pointed in the wrong direction'. Nelson and Winter (1982) argued that since organisational routines are delivered through performances, the routine should respond to the performance feedback: if the feedback received after the execution of a routine indicates that the performance is no longer satisfactory, the organisation should initiate a 'routine-guided, routine-changing' process. Cyert and March (1992) studied performance feedback as the mechanism through which an organisation learns about the appropriateness of the performance of its service which can help trigger a change in the routine process.

A broad definition of organisational routine as processes of the organisation whose complexity vary from simple operational routines to revisions in the corporate strategy was given by Nelson and Winter (1982). They noted that the organisation's structure routines could be conceived as a hierarchy with some routines playing an operational role and others performing the organising function. According to Pavlov and Bourne (2011), given the fact that an organisation's routines determine what the organisation is about and what it is measured for, the attitude and behaviour of the staff should embrace it.

## 4.9 Payment Mechanisms as Part of PFI Contracts

Payment mechanisms and performance monitoring are an integral part of a PFI contract. A good payment mechanism reflects significant potential variation in payment arising from the quality and quantity of the service provided. While it is accepted that within a service contract, performance failures are inevitable consequences, it is important that these are highlighted and reviewed to ensure that they are minimised and that the customers receive the desired service (Herbert and Jon 2009). A lack of robust performance monitoring could mean that poor standards of performance go undetected, leading to negative experiences for end users and potential reputational issues for the Trust and the service provider. If performance standards are not measured in accordance with the performance measurement criteria, this could result in non-compliance with the contract agreement. Inaccurate reporting of performance standards could also have an adverse effect on the payment mechanism and therefore overall income to the service provider. One desirable aspect of the PFI is the introduction of a payment for performance incentive into areas of public-service delivery where they did not exist previously. However, Berg et al. (2002) argued that the extent to which incentive payment occurs may be a function of performance. For example, failure to specify that all of the equipment in a PFI hospital life cycling contract must work when it was migrated to the NHS Trust would give the contractor a strong incentive to economise by not spending money getting the equipment operational (Berg et al. 2002).

The Institute of Medicine (2001) recommends that effective delivery systems should align financial incentives with the implementation of healthcare services based on best practices and the achievement of better patient outcomes. 'Substantial improvements in quality are most likely to be obtained when service providers are highly motivated and rewarded for carefully designing and fine-tuning healthcare processes to achieve increasingly higher levels of safety, effectiveness, patient-centredness timeliness, efficiency and equity' (Institute of Medicine 2001). The National Audit Office (2010) unequivocally states that client management can make unitary financial deductions from the service providers for failure to meet performance indicators and if parts of a PFI building is not available for use for any reason. The NAO (2005) document also encourages client management to use financial deductions to penalise poor performance to encourage improvement.

The payment mechanism calculates and measures performance of each service (excluding the managed equipment service and the telephone service) in each contract month against a defined set of performance indicators (PIs). Each of the PIs is scored out of 20 in accordance with the preset calibrations which correlate the scores with the number of failures, incidents and satisfactory responses from customer surveys. The points scored in respect of each PI are totalled together through a weighting system. The sum of the weighted scores for each PI relating to a specific service provides an overall score out of 20 for the service. The sum of the weighted scores for each individual service provides an overall score out of 20 for all the FM services together. Unavailability deductions are equally made to the extent that service suffers in respect of an availability unit; a further deduction in respect of a failure to provide the services to the availability unit is also made. Performance indicators and service weightings are reviewed annually by the parties, three months prior to the start of the next contract year, with a view to confirming or revising the existing arrangements.

## 4.10 Performance Monitoring Tools

Various means of performance measurement tools are considered but the role of key performance indicators (KPIs) will be reviewed in the context of this research.

### 4.10.1 The Service-Level Agreements (SLAs)

In addition to the previous reasons given for measuring the delivery of FM services, the standard of service that is expected from a service provider in a PFI project is specified in a service-level agreement (SLA) developed as part of the contractual agreement. The standards bound the service provider to perform and measure services against the outputs specified in the contract agreement. According to 4Ps (2005), the function of the contract management is to assess the performance of the service provider against the contract specifications by ensuring that:

1. Best value is realised.
2. Change is monitored.
3. Risk is managed.
4. Service improvements are implemented when service standards are not fulfilled.
5. Remedial measures being implemented are effective.
6. Monitoring meetings are held on a regular basis and in accordance with the contractual requirements.
7. Obligations for payment are made on time and in accordance with the contract.
8. Continual improvement is encouraged.

It is a requirement for PFI contractors to also report their performances against a set of PIs which together form a PM system. For example, record of a reported faulty bulb, when it was reported, mended, and if within the time specified in the contract. Until there is a change in regulation, measuring performance would remain a necessary condition for managing performance in a PFI facilities management. The service-level agreement specifies the contractual obligation that is expected of the service provider. Atkin and Brooks (2009) described SLA as a document specifying the expected level of services and the quality of performance expected which are key aspects of facilities management service. These formal documents emphasise the standards of service expected rather than the processes involved and contain features to facilitate and promote the development of good working relationships between both parties. According to 4Ps (2005), a well-drafted output specification agreement is fundamental to the successful delivery of a PFI service contract.

These agreements are needed to measure the delivery of FM services whether retained in-house or outsourced. Performance measurement is also entrenched in the terms of agreement, which contains the critical success factors and key performance indicators and ensures that performance aims are in alignment with the organisation's business objectives (Atkin and Brooks 2009). The SLA in most cases includes some performance targets as incentive to the service provider to deliver the services (4PS 2005).

### 4.10.2 The Balanced Score Card (BSC)

The continued increase in customer expectations, service competition, the desire to make efficiency savings and improve processes have all contributed to the push for the use of performance measurement in the healthcare sector. Service providers can no longer ignore the importance of strategy execution over strategy definition in attaining sustainable competitive advantage. To be competitive in any business, a service provider must 'select the right measures, define stretch goals, track and measure both results and the process for attaining the results, and attach meaningful consequences to performance' (Schneier et al. 1995).

The balanced scorecard provides management with a comprehensive framework that translates an organisation's strategic objectives into an intelligible set of performance measures and works on the premise to motivate breakthrough improvements in critical areas, such as service, product, process, customer and learning development. According to Schneier et al. (1995), the balance scorecard has four perspectives: the financial indicators with measures of performance for customers, the internal processes, innovation and improvement activities. It is evidenced from the operations of the healthcare services that business can no longer service on financial indicators only and the balanced scorecard has four key elements of which one is the 'customer'.

Amaratunga and Baldry (2000a) state that the balanced scorecard is 'a conceptual framework for translating the organisation's vision into a set of performance indicators'. The NHS Institute for Innovation and Improvement (INI 2008) acknowledged the fact that it would require a major piece of organisational change to produce a balanced scorecard for an organisation as complex as the health Trust whose activities are myriad and intertwined. Therefore, cascading balanced scorecards to the various departmental levels of the organisation would dominate the activities of most of the managers.

Spitzer (2007) opined that the popularity of the balanced scorecard was necessitated by four innovative principles – the financial perspective describes how the establishment would sustain the value of the shareholder in terms of profit and revenue growth; the customer perspective interprets customer satisfaction, increased customer perception of value and improved brand image; the internal process interprets how the organisation intends to improve service quality and customer relationship management; while learning and growth focuses on the effort of the organisation to increase the skill of the workforce through training and development and increased innovation.

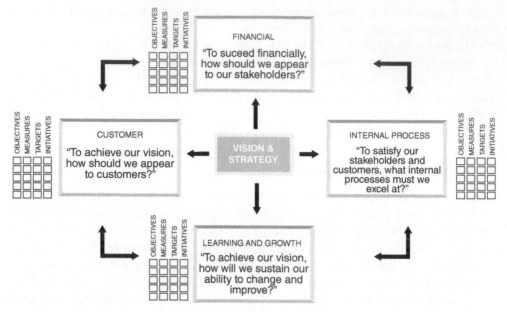

**Figure 4.1** The Balanced Scorecard. *Source*: NHS Institute of Innovation and Improvement 2013.

The aim of the balanced scorecard is to develop specific performance measures in relation to four areas, as shown in Figure 4.1. The financial perspective as the lead element helps facilitate the strategic financial objectives of the other three elements by identifying and measuring their various objectives. In perspective, customer satisfaction enhances financial success, good business process enhances customer satisfaction, while continuous improvement practice ensures a good organisational performance.

### 4.10.3 The Business Excellence Model (BEM)

The Business Excellence Model (BEM) is a measurement tool for measuring simultaneously employer, customer and shareholder satisfaction within the organisation. It measures how well different parts of the organisations are performing in direct proportion across each area, and is also able to compare the same business in different places (Gopal 1998). The model, which was developed in 1990 by the European Foundation of Quality Management (EFQM), is practical, non-prescriptive and operates on a nine operational criteria: customer satisfaction, financial results, people satisfaction, leadership, policy, resources and strategy, processes management, people management and impact on society, achieved through acting on enablers (Meng and Minogue 2011). The criteria also describe the difference between the cause and effect relationship and the enabler's results of the business processes within the organisation (Turner 2008).

## 4.10.4  Benchmarking

Benchmarking is a form of performance measurement that seeks improvements by comparing a provider's performance level to the benchmark of another or the best in class (Assessment and Qualifications Alliance (AQA) 2006). The British Institute of Facilities Management (BIFM 2013) provided that benchmarking can help drive an organisation's FM performance strategy if used in conjunction with other measurement tools. The organisation states that 'benchmarking is a business "MOT" that measures whether you are running as efficiently as your competitors' (BIFM 2013). The BIFM stated that comparisons against benchmarks capture the notion of efficiency, with each part of the system compared to a benchmark, to identify where area of improvement can be made. AQA (2006) stated that benchmarks 'should reflect the best current assessment of optimal care and efficiency rather than average performance, wherever possible'. The BIFM (2003) identified barriers that can militate against a successful benchmarking process as: amount of time – the amount of time and effort it takes to generate benchmarking reports; confidentiality – the fears of sharing data and how to make it somehow unspecified; the need for the industry to embrace a collaborative process; and using benefits realised for decision making.

An organisation can use its performance measurement result to benchmark its services against a similar organisation's, which can provide inputs for developing targets and learning best practices.Benchmarking is a performance measurement model that seeks to push performance improvement by comparing a service provider's level of performance to the benchmark of similar service elsewhere (Amaratunga and Baldry 2002). Benchmarking according to Camp (1998) is 'the search for industry best practice that leads to superior performance'. The author added that the process of benchmarking does not only provide a way of ensuring if the organisation is satisfying the customer, but as a monitoring mechanism that adjusts to customer requirements as their taste and demands changes.

Benchmarking within hospitals facilities management services can be an effective tool in determining how well a service is performing in comparison to another hospital facility and can provide improvement opportunities, especially when applied from the perspective of the customer who is the reason for the component exercise. The process can either be internal or external with the latter having the determinant strength to help improve processes (Tucker and Pitt 2009). A number of tools exist within the NHS to improve service performance, but none, however, is more powerful than benchmarking, which compares one asset to another and can be redesigned to make as a cost-saving exercise (Ellis et al. 2010).

However, benchmarking in the NHS still remains problematic with every sector challenging detailed comparison and analysis on a number of variant performance criteria. The problem with benchmarking is that most of the benchmarks are derived on the basis of a causal inference method and are highly susceptible to false linkages and conclusions. Benchmarks 'should reflect the best current assessment of optimal care and efficiency rather than average performance, wherever possible' (Trisolini et al. 2006). According to Silver (2012), one of the problems with benchmarking is that it is wholly retrospective: it looks back at what had been done, which is not a merit on its own.

A good benchmarking should demonstrate more than what is replicated, but the process only improves in relation to a selected benchmark data used for comparison. There is also the jeopardy of believing that the data used for the benchmarks are accurate or relevant to the organisation; rather, it may be the case that the data used for the benchmarking are the ones that were easily accessible rather than what is relevant to the object of the exercise (Silver 2012).

As an approach to this research, the CQC (2013) audits surveys results of 2013 of various hospitals in England was used to compare the FM services at the Central Manchester University PFI hospitals as a benchmark to other hospitals to know how the services compares on the same survey feedback questions. The findings are discussed in Chapter 5.

### 4.10.5 Patient-Led Assessment of Care Environment (PLACE)

The condition of wards and their cleanliness has a huge effect on how comfortable, relaxed and confident patients feel, which in turn affects the recovery process of a patient. The PLACE assessments provide a view of how the NHS is performing against a range of non-clinical issues which can have a negative impact on the overall patient experience – appearance and maintenance of healthcare premises, cleanliness, the building conditions, the extent to which the environment supports the delivery of care with privacy and dignity, and the quality and availability of food and drink. The PLACE audit is a further shift towards patient power to judge the service in which they are the end users. A crucial component of the assessment process is the involvement of patients as visual assessors of the services, covering the people who experience the hospital as a customer rather than a provider, and so encompasses relatives, friends and advocates (Department of Health 2013).

### 4.10.6 The Matron's Charter

Nurses and midwives are looked to by the general public to ensure that the hospital environment is clean and safe for the patient. Their leadership is essential, but they cannot succeed alone. It was, therefore, imperative for the NHS authority in 2003 to recognise the role of the people that deliver the service to help improve the quality of the service by vesting the responsibility of monitoring the standards of cleaning on hospital wards with the matrons. The annual 'cleanliness' programme of audits by the hospital matrons is to ensure that every area receive the desired services. The charter, which has ten guiding principles for delivering cleaner hospitals, is aimed at the hospital staff in conjunction with the end users whose opinion for improvement are most relevant (NHS 2004).

The patient environment will be well-maintained, clean and safe:

1. Sufficient resources will be dedicated to keeping hospitals clean: keeping the NHS clean is everybody's responsibility.
2. Establishment of clean culture across the hospital unit should be introduced by matrons.
3. While matrons have the authority to withhold payment, the infection control team and nurses should be involved in creating cleaning contracts.

4. Cleaning routines will be clear, agreed and well-publicised.
5. Cleaning staff will be recognised for the important work they do. Matrons will ensure sure that they feel part of the ward team.
6. Patients will have a part to play in monitoring and reporting on standards of cleanliness.
7. Specific roles and responsibilities for cleaning will be clear.
8. Educating all healthcare staff in infection control should be mandatory.

### 4.10.7   The National Standard of Cleanliness

The National Standards of Cleanliness for NHS Trusts provide a toolkit for auditing and assessing the performance of the service measured against specific cleaning outcome requirements. A core standard in standards for better health is the provision of a clean and safe environment for the well-being of the patients. Hospitals recognise the role cleaning has in ensuring that the risk to patients from healthcare associated infections are minimal. Increased public concern about healthcare associated infections means that a hospital needs not only to be clean, but also able to demonstrate how and to what standard they are kept. The hygiene practice code bill for the prevention and control of healthcare associated infections have further placed the onus and responsibility on various NHS trusts to ensure that cleaning services are sufficiently resourced and defined with a strategic cleaning specification in the form of schedules and cleaning frequencies. According to the NHS National Patient Safety Agency (2007) the most important thing is for the hospital to be clean and should remain the focus of a hospital's cleaning facility, but it should not be at the expense of the service delivery.

### 4.10.8   KPIs as Instrument for Measuring Effective Delivery of FM Services

A key performance indicator (KPI) is a form of measurement used to evaluate the success of a particular activity in an organisation. The implementation of the correct KPIs rely on the premise of understanding the key objectives of the organisation and the customers it serves. The use of effective KPIs as a measurement parameter for every process of services should be tailored to the needs of the consumer, who essentially is the final recipient and better placed to assess the quality of service rendered. This includes the patient and the staff delivering the service, the quality of the service to be delivered, patient safety, quality of care, outcomes and so on.

According to Cole (2007), hospital designs should meet the needs of six categories of people: the patients, staff, hospital management, facilities management, the wider health system and the general public. Given the complexity of the healthcare system, many valid and important objectives may be inherently difficult to measure; relating to a quote attributed to Albert Einstein that 'not everything that can be counted counts and not everything that counts can be counted'. Neely et al. (1995) viewed performance measurement as 'the process of quantifying the efficiency and effectiveness of action', and as 'a metric used to quantify the efficiency and/or effectiveness of an action'.

Marshall et al. (1999) describe PM as '... the development of indicators and collection of data to describe, report on and analyse performance'. Neely et al. (1995) described performance measurement as the process of quantifying action, and more specifically define it as 'the process of quantifying the efficiency and effectiveness of action'.

Chan and Chan (2004) defined KPIs as 'key requirements that focus on critical aspects of output or outcomes; they help identify what stakeholders view as being important in meeting their needs in the overall delivery of a building project'. Chan and Chan (2004) also state that for a performance measurement indicator to be effective, it must be accepted, effective, feasible and understood. It was identified by Vokurka and Fliedner (1995) that the holistic and strategically oriented modelling of critical success factors is attributed to the non-financial qualitative performance support. They also identified that since no single measure can make adequate provision for a clear performance target, it will be desirable to establish a balance between the financial and non-financial measures.

Furthermore, Slater et al. (1997) qualified this perspective by suggesting that there should be a limit of between 7 and 12 PIs in order to assess the framework. The approach by Becker (1990) to FM performance assessment stressed the need for indicators that would be relevant to the business-oriented customers to support comparisons between the desired and actual FM performance. He highlighted the importance of relating measures to strategic objectives and how they must be dynamic and easy to communicate, and how well FM services were performing and where improvement was needed. Three key performance criteria for FM of quality, cost and delivery were identified by Varcoe (1993) as the primacy of added value in the business process. Varcoe (1998) in a later study also noted that any inherent limitations of using financial measures such as cost-per-capita or costs-per-area must be relevant to the business context. While supporting the recommendation of Slater et al. (1997) that the number of PIs should be minimised to allow for effective management, Varcoe (1998) advised that it is usually enough to have five or six well-defined objectives, each with four to six key facilities PIs. Wang (1992) observed that the use of PIs had become an exception rather than a norm; with lack of agreement across the practice in relation to the nature and terminology for PIs, other observations about the impracticality of creating a set of PIs that will be generally acceptable may be endorsed. Gearey (1996) provided a distinction between effectiveness, efficiency, and flexibility which can influence the concept of performance measurement tools. Effectiveness and efficiency are different concepts. This difference is frequently overlooked in interpreting measures of the performance of services. It is one thing to say that a service has produced positive outcomes with respect to one or more of its several stated objectives; it is another to say that it is achieving these outcomes in the most efficient manner relative to best uses.

Several measurement tools have been developed to monitor different aspects of the healthcare facilities management services. During the process of developing measures that can help manage the complexities of hospital facilities, Pullen et al. (2000) identified the principal performance issues in the hospital as income and operating cost; asset value; occupancy; and size of the hospital. The authors claimed that the identifying factors led to the creation of several KPIs, with four directly relating to hospital revenue. The quest for accountability, higher performance and cost efficiency has necessitated the need to obtain the best level of resources to achieve the desired objective (Shohet and Lavy 2004).

'Key performance indicator' is defined by Parmenter (2010) as a set of measures focusing on those aspects of organisational performance that are the most critical for the current and future success of the organisation. The author provided four types of KPI commonly in use:

1. Key result indicators (KRIs), which inform the organisation of its performance in relation to set objectives or critical success factor.
2. Result indicators (RIs) inform the organisation about its performance.
3. Performance indicators (PIs) inform the organisation what to do.
4. Key performance indicators (KPIs), which inform the organisation about what it has to do to increase performance dramatically.

According to Parmenter (2010), there are seven characteristics of KPIs:

1. They are non-financial measures, i.e., not expressed in monetary form, e.g., pounds, dollars, euros, etc.
2. They are frequently of measurement (daily, weekly or monthly).
3. They are acted on by the top management.
4. They clearly stipulate the action required by staff to understand the measures and the remedies.
5. They are measures that tie responsibility down to a team, e.g., management team delegating team leader to take the necessary action.
6. They have a significant effect on one or more of the critical success factors and more than one perspective.
7. They encourage necessary action to ensure that the process have been tested to ensure success and to avoid dysfunctional behaviour.

The BIFM stated that effective performance measurement has increasingly become reliant on KPIs, which reflect the prominent success factors against given objectives. BIFM provided that many KPIs are varied and are mostly composed by individual teams or organisations to suit a purpose while offering a generic template.

While Kaplan and Norton (1996) recommend not more than 20 key performance indicators, Hope and Fraser (2003) suggest fewer than 10 KPIs (cited in Parmenter 2010). That is, there are about 10 KRIs which tell the organisation how it has done in an objective; 80 result and performance indicators tell the organisation what it has done with a direction on what to do; and 10 KPIs tell the organisation what to do in order to dramatically increase performance (Parmenter 2010). The author opined that in most cases, it is better to have fewer measures as opposed to more than necessary.

Drawing on work by TRADE (1995), the set of indicators are grouped into six domains:

- Effectiveness, which is the characteristic of a process that shows the extent to which the process output conforms to requirements;
- Efficiency, the characteristic which shows the degree to which the process produces the required outcome at minimum cost;
- Quality, the extent to which the service meets requirements and expectations of the customer;

- Timeliness, the accomplishment of a requirements relating to the time it takes to complete a process, which must be established and defined to know what constitutes timeliness for a given process; and
- Productivity, the value adding process which increases the outcome of the service.
- *Safety*: Protecting staff and customers from any risk or harm while in the environment of the organisation.

In lean management, a process is not necessary if it is does not add value to the service delivered; safety protects the employees and the service users of the organisation against unnecessary harm or risk while in the hospital environment.

## 4.11   The Importance of the Helpdesk for the Success of Maintenance and Facilities Management Services in PFI Initiatives

The facilities management service help desk on any PFI scheme should be the centre of any communication strategy, enabling service users to log requests for service through a single point of contact. The help desk should operate using a computerised logging system, in which predetermined priority levels, as described within an organisation's performance standard schedules for tasks set and jobs allocated, together with response times and task completion times. The help desk is the first point of contact for all service-related calls and requests for additional service. The help desk accepts requests varying from the changing of a light bulb to the full infection clean of a ward. The service monitors the response time of all tasks and provides data for monitoring purposes. The system should also have the ability to track resource allocation and analyse response times in line with service agreements.

## 4.12   The Performance Monitoring Process

There should ideally be an independent consultancy body tasked to monitor the performance of each maintenance and facilities service as part of a PFI contract and to ensure that each service is adequately monitored with records maintained. The performance monitoring system should be outlined in the SLA of the contract specifications. It requires that for the duration of the contract, the service provider shall provide to the client organisation a demonstration of the technology-based monitoring system, utilise and develop all checklists and supporting documentation which govern the monitoring of each of the services provided by the contracting company.

The performance monitoring process should incorporate a payment mechanism system, which calculates the financial deductions, and the service failure points on a monthly basis, based on the data input from the monitoring. The independent monitoring team should collaborate with the client body to resolve quality and performance issues, ensuring objective and transparent contract monitoring and adherence to the performance standards. Each performance standard is monitored in accordance with the performance measurement criteria, the monitoring method statement and the monitoring frequency. Additionally, each performance standard has a performance factor which determines its severity. An example of the different performance factors is shown in Table 4.1.

All the services provided by the service provider have a service-level specification and a performance standard which the company must adhere to. With regards to maintenance services, these could include compliance measures and regular statutory inspections and certification for such aspects as legionella and asbestos management, alongside electrical, gas and other mechanical and electrical installations. In addition, as part of maintenance responsibilities, the service provider is normally required to provide a scheduled and reactive cleaning service on a daily basis to meet the requirements of the client organisation in all areas, in accordance with the service-level specification and the performance standards. Performance reports are normally submitted monthly as a requirement of the contract schedule by the service provider, reporting the level of compliance for each month for both the soft and hard FM services, including all maintenance aspects. Each of the services provided by the service provider should have a set of defined performance standards against which it is monitored. This details a number of checks that the company is required to undertake, each of which is called a performance indicator (PI). Each PI has an agreed way of being monitored, which is the Monitoring Method Statement (MMS). The scoring, called the Performance Measurement Criteria (PMC), is detailed on the Performance Monitoring System (PMS) on the contract specification schedule. The

Table 4.1 Performance monitoring factors.

| Performance monitoring scores | | |
|---|---|---|
| Performance Factor (PF) | Description | Performance Weighting Factor (PFW) |
| PF 9 | **Mandatory** – e.g., legislative or otherwise mandatory requirements which must be achieved by Project Co at all times. | 9 |
| PF 6 | **Critical** – e.g., the delivery of the performance standard is critical to the clinical services and/or the health safety and welfare of the trust's patients, staff and visitors. | 6 |
| PF 4 | **Important** – e.g., the delivery of the performance standard is important to the continuity and quality of the clinical services. | 4 |
| PF 1 | **Routine** – e.g., the delivery of the performance standard does not directly affect the continuity or delivery of the clinical services; however, if not corrected, could have a cumulative effect on service integrity. | 1 |

*Source*: The service provider, CMUH.

frequency with which each PI is monitored is also specified in the Monitoring Frequency (MF), alongside the particular Scoring Method (SM) frequencies which range from daily to biannually.

An independent, full-time monitoring team set up by the service provider and the client body should have overall control and charge of the monitoring, which was previously completed by supervisors. The client body (acting by itself or through the appointment of a subcontractor) may, at its own expense, also carry out such monitoring of the service provider's performance and monitoring of the services when required. A system is used to generate checklists for each service in relation to the individual's PI, in accordance with the MF. Once the necessary check has been completed, the findings are then recorded on the checklist and the overall scores calculated.

### 4.12.1 The Monitoring Methodology

The service provider would normally use an electronic monitoring system to generate the checklist for the daily monitoring and to record the performance of all services. It should contain the checklists completed for each PI, as well as additional information such as details of all rooms on site, including the description and room number. The help desk team also can utilise this system to log information from calls they receive in relation to ad hoc requests, for all the services including any complaints. The monitoring system can also be used to generate reports for each service to show overall performance for a specified period of time, as well as calculating financial deductions for failed checklists. In addition, there can also be a joint monitoring arrangement with both client and service provider carrying out joint monitoring activities in accordance with the agreement. The routine joint monitoring forms can sometimes deal with 10% of the monitoring activities undertaken by in accordance with the agreement for each contract month.

### 4.12.2 Evaluating the Impact of KPIs on the Effective Delivery of FM Services

Measurement evokes uneasiness, anxiety and frustration among the people concerned: the service provider, the client management, the staff whose work are being measured, the service user, the staff that are doing the monitoring and even the people who are seeking the data for different purposes (Loeb 2004). As Loeb (2004) revealed, there is a distinct need to improve the overall performance necessitated by the increasing demands by clients and their end users for demonstrable evidence of quality, and demands for accountability, which have become a major driver. For this reason, maintenance and facilities services need to be constantly reviewed in light of changing requirements, service standards and lessons learned.

The most significant barriers identified by using performance measurement are the issues of having too many indicators or not having the right information to aid decision. Hibbard et al. (1997) suggested that service providers are so overwhelmed by the amount of performance information at their disposal that it now acts as a deterrent. Service providers contend with as many as three different categories of PIs (e.g., response times, service quality and consumer satisfaction) each with multiple measures. However, in some cases there may be significantly more performance indicators which can sometimes be unwieldy to manage. In these situations, not only is the amount of information prohibitive to making decisions, but managers may also find it difficult to assimilate all of the variables into measures from which they can make decisions. There is also the burden of other measurement systems that present difficulties in comparing information across services, including some developed by the service provider to reflect their own quality standards. The different approaches might differ by definitions, time periods, measuring methods or adjustment factors. The lack of standardised measurements adds cost to the services to meet the needs of the different requirements, and also limits the ability to compare measures across measurement systems (Eddy 1998; Maxwell et al. 1998).

The other common limitation of performance measurement is the inability of the service provider to measure information that users really want. In general, cost is among the most important piece of information in the provision of patient services in the healthcare services.

Referring to the work of Leatherman and Sutherland (2008) and Parmenter (2010) the main set of performance indicators, as mentioned in section 4.10.8, are grouped into six domains for audit purposes:

*Effectiveness*: A characteristic in the process indicating the degree to which the process output conforms to requirements of the service user. Effectiveness and efficiency are different concepts. This difference is frequently overlooked in interpreting measures of the performance of services. It is one thing to say that a service has produced positive outcomes with respect to one or more of its several stated objectives; it is another to say that it is achieving the outcomes in the most efficient manner relative to best uses. Efficiency is only achieved by a clear definition of responsibilities, task training, and a committed approach to continuous quality improvement.

*Efficiency*: The ultimate aim of designing a service is to ensure that the end user is satisfied with it. Therefore, the test and the evaluation process should lie with the consumer whose needs must be taken into consideration and translated into measurable characteristics (Spitzer 2007) thus ensuring patients receive care in an environment that is clean, safe and welcoming. This is achieved by robust monitoring procedures, acting on client feedback, providing services which are customer focused by introducing a comprehensive customer care training programme and setting time for tasks to meet client requirements and demands, and operating as an integral part of the client team, minimising the risk of non-compliance and poor performance while adopting best practices methodologies.

*Quality*: The degree of satisfaction achieved by a customer for using a service. A characteristic in a process indicating the degree to which the process produces the required output at minimum resource cost. Apparently, performance should be measured from the point of view of the customer not the providers.

*Timeliness*: Measures whether a unit of work was done correctly and on time. Criteria must be established to define what constitutes timeliness for a given unit of work. The criterion is usually based on customers' requirements.

*Productivity*: Synergy – the difference in the value adding process by the value of the labour cost.

*Safety*: Protecting staff and customers from any risk or harm while in the environment of the organisation.

## 4.13    Key Issues Arising for Performance Management as Part of a Maintenance Management and FM Tool on PFI Schemes

Clearly, performance measurement has a powerful impact on the processes of an organisation, but its effect is far from predictable and is poorly understood (Pavlov and Bourne 2011). According to Pavlov and Bourne, the antidote to the problem is to discover the contents and mechanisms of the 'black box' that sits between performance measurement and the outcome which contains the organisational processes that delivers performance. While proving that the central function of any performance measurement process is to provide regular and valid data indicators of the performance outcomes, Sachs (2011) opined that performance measurement should not be limited to data on outcome and efficiency indicators alone; rather, it should include information that can help measure the effectiveness of the process and gain insight into the causes of the outcomes and the relative costs of the service in proportion to the outcome it produces.

A performance measurement is only as good as the outcome it tracks, with each of the processes needing a specific list of outcomes that is important to it. Selecting the specific indicator to measure is a key part to developing a performance measurement system. However, a lot of organisations base their selection of indicators on how readily available the data are and not how important the data are for measuring the achievement of the outcome (Hatry 2006). A major potential criticism of performance measurement from a facilities and maintenance management perspective is that it focuses attention on the indicators to the neglect of outcomes that cannot be measured. For the system to be comprehensive, it needs to include a set of indicators that track other outcomes. For example, a law enforcement agency that focuses solely on number of police arrests will tempt staff to 'harass' individual citizens in order to increase the desired values. Including other indicators, such as the number of complaints validated, will considerably reduce the problems and alter the incentives. Usually, more than one indicator will be appropriate for measuring an outcome. Seeking comprehensiveness will be constrained by available measurement resources and inherent data problems (Hatry 2006).

## 4.14    Conclusions and Reflections

Performance measurement is not an end in itself, but a valuable tool that can help improve the delivery of maintenance and other FM services for the benefits and well-being of the service user. Performance measurement is well established, especially in the health sector, purely to measure what is being achieved in terms of value for money and as an indicator of organisational efficiency. Furthermore, given the amount of performance indicators inundating the service provider, it is essential to focus attention and resources on those performance measurements that drive improvement. The numerous overlapping measurement requirements faced by the service provider thus create a need to harmonise the situation in order to reduce cost and the administrative burden, and home in on the most important measures which focus on improving the maintenance and facilities management services.

Several factors were identified as contributors to the slow growth of performance measurement in a PFI context especially. Despite the proliferations of quality dashboards, scorecards and the enormous advances in the technological aspect including human progress, performance measurement, for most part, still makes people feel helpless rather than empowered. The negative perception of measurement has not changed much; neither has the experience of the employees. There is a stereotypical notion that measurement, at best, is a necessary evil, a compliance not a commitment, driven only by top management for reverence reasons. When staff are asked about their personal measurement experience, the obvious enmity response suggests that performance measurement may be perceived with negativity. There notion of fear and judgement associated with performance measurement can also put negative pressures on people and sometimes propel them to do whatever it takes in order to comply with the expectations of performance measurement. As Edward Deming explained, even if it means attaining a particular score to avoid failure, people will do whatever it takes to meet targets 'even if they have to destroy the organisation to do so'. The tension between using measurement to enable learning and improvement and using measurement to drive payment mechanism also have some underlying implications. When measurement is heavily linked to the payment mechanism, service providers may limit their focus to limited actions that affect specific measures as opposed to building coalitions and collaborations that address the broader purpose.

## Acknowledgement

This chapter has been largely based on the research of Ben Lyere, who previously has carried out extensive research into the challenges of performance measurement of facilities and maintenance management in a PFI hospital context.

## References

4Ps (2005). *4ps review of operational PFI and PPP projects*. November. http://test.4ps.gov.uk/UserFiles/File/Publications/review_of%20_operational_PFI_PPP_schemes.pdf (accessed 17 October 2013).

Amaratunga, D. and Baldry, D. (2000a). Assessment of facilities management performance in higher education properties. *Facilities* (7/8): 293–301.

Amaratunga, D. and Baldry, D. (2000b). Building performance evaluation in higher education properties. *Facilities* 18 (7/8): 293–301.

Amaratunga, D. and Baldry, D. (2002a). Balanced scorecard: universal solution to facilities management? In: *Proceedings of the Euro FM Research Symposium in Facilities Management* (ed. K. Alexander). Salford: The University of Salford.

Amaratunga, D. and Baldry, D. (2002b). Moving from performance measurement to PM. *Facilities* 20 (5/6): 217–233.

Amaratunga, D. and Baldry, D. (2003). A conceptual framework to measure facilities management performance. *Property Management* 21 (2): 171–189.

AQA (Association of Quality Assurance) (2006). http://www.aqaalliance.org/files/ PrinciplesofEfficiencyMeasurementApril2006.doc (accessed 18 November 2013).

Arah, O.A., Klazinga, N.S., Delnoij, D.M., et al. (2003). Conceptual frameworks for health systems performance: a quest for effectiveness, quality, and improvement. *International Journal for Quality in Health Care* 15 (5): 377–398.

Armstrong, M. and Baron, A. (2009). *Managing Performance: PM in Action*. London, UK: Chartered Institute of Personnel and Development.

Atkin, B. and Brooks, A. (2009). *Total Facilities Management*, 3e. Oxford: Blackwell Science.

Barret, P. (2000). Achieving strategic facilities management through strong relationship. *Facilities* 18 (10/11/12): 421–426.

Barrett, P.S. and Baldry, D. (2007). *Facilities Management: Towards Best Practice*, 2e. Oxford: Blackwell Science.

Becker, F. (1990). *The Total Workplace: FM and the Elastic Organisation*. Chapter 14. New York, NY: Van Nostrand Reinhold, 291–305.

Berg, S.V., Pollitt, M.G., and Tsuji, M. (2002). *Private Initiatives in Infrastructure: Priorities, Incentives, and Performance*. Cheltenham, UK: Edward Elgar Publishing.

Berwick, D.M., Nolan, T.W., and Whittington, J. (2008). The triple aim: care, health, and cost. *Health Affairs* (Millwood) 27 (3): 759–769.

Bititcti, U.S., Turner, T., and Begemann, C. (2000). Dynamics of performance measurement systems. *International Journal of Operations & Production Management* 20 (6): 692–704.

British Institute of Facilities Management BIFM (2003). What is facilities management? http://www.bifm.org.uk/bifm/home (accessed 17 August 2012).

British Institute of Facilities Management BIFM (2013). *Benchmarking: effective PM for FM*. *FM Leaders Forum*. Discussion Paper June 2013.

Camp, R.C. (1998). *Benchmarking: The Search for Industry Best Practices that Lead to Superior Performance*. Milwaukee, MI: Quality Press, 44–47.

Chan, A.P.C. and Chan, A.P.L. (2004). Key performance indicators for measuring construction success. *Benchmarking* 11 (2): 203–221.

Chandrupatla, T.R. (2009). *Quality and Reliability in Engineering*. Cambridge and Oxford: Cambridge University Press.

Cole, J. (2007). *Achieving the client objectives in PFI projects*. www.dhsspsni.gov.uk/pfi_kells_ part_2.pdf (accessed 28 October 2013).

CQC (Care Quality Commission) (2013). University Hospitals of Morecambe Bay NHS Foundation Trust; Royal Lancaster Infirmary Furness General Hospital: Investigation follow-up report.

Cyert, R.M. and March, J.G. (1992). *A Behavioural Theory of the Firm*, 2e. Cambridge, MA: Blackwell.

Davies, H.T.O. and Lampel, J. (1998). Trust in performance indicators? *Quality in Health Care* 7: 159–162.

Dennis, S.O. (1995). Joint Commission on Accreditation of Healthcare Organizations. Measurement and accountability: taking careful aim. *Journal of Quality Improvement* 21 (July): 354–357.

Department of Justice (2004). *Improving healthcare: a dose of competition. A report by the Federal Trade commission and the Department of Justice July 2004*. http://www.ftc.gov/reports/healthcare/040723healthcarerpt.pdf (accessed 29 August 2013).

Dixon, T., Pottinger, G., and Jordan, A. (2005). Lesson from the private finance initiative in the UK: Benefits, problems and critical success factors. *Journal of Property Investment and Finance* 23 (5): 412–413.

DOH (2010). *Equity and Excellence: Liberating the NHS*. UK: The Stationery Office Limited on behalf of the Controller of Her Majesty's Stationery Office.

Drucker, P.F. (1968). *The Age of Discontinuity Guidelines to Our Changing Society*. London, UK: Transaction Publishers New Brunswick.

Drucker, P.F. (1998). *The Coming of New Organisation, Harvard Business Review on Knowledge Management*. Boston: Harvard Business School Press.

Eddy, D.M. (1998). Performance measurement: problems and solutions. *Health Affairs* (Millwood) 1998; 17 (4): 7–25.

Elg, M., Broryd, K.P., and Kollberg, B. (2013). Performance measurement to drive improvements in healthcare practice. *International Journal of Operations & Production Management* 33 (11/12): 1623–1651.

Ellis, C., Baldwin, E., and Dick, R. (2010). Radical 10-point plan to refocus NHS estate. *Health Estate* 2010 Nov; 64 (10): 47–52.

Federal commission and the Department of Justice (2004). *Improving healthcare: a dose of competition*. http://www.ftc.gov/reports/healthcare/040723healthcarerpt.pdf (accessed 18 November 2013).

Fitzgerald, L., Johnston, R., Bignall, S., et al. (1993). *Performance Measurement in Service Businesses*. Chartered Institute of Management Accountants.

Froud, J. and Shaoul, J. (2001). Appraising and evaluating PFI for NHS hospitals. *Financial Accountability and Management* 17 (3): 247–270.

Gearey, D. (1996). Managing facilities performance. Facilities Management (October), pp. 7–9.

Gelnay, B. (2002). Facility management and the design of Victoria Public Hospitals. *Proceedings of the CIB Working Commission 70: Facilities Management and Maintenance Global Symposium 2002*. Glasgow, 525–545.

Goodrich, J. and Cornwell, J. (2008). Seeing the Person in the Patient. The Point of Care Review Paper. The King's Fund. http://www.kingsfund.org.uk/sites/files/kf/Seeing-the-person-in-the-patient-The-Point-of-Care-review-paper-Goodrich-Cornwell-Kings-Fund-December-2008.pdf (accessed 18 November 2013).

Gopal, K. and Patrícia, M. (2003). Sustaining healthcare excellence through performance measurement. *Total Quality Management & Business Excellence* 14 (3): 269–289.

Gopal, K.K. (1998). Measurement of business excellence. *Total Quality Management* 9 (7): 633–643.

Hatry, H.P. (2006). *Performance Measurement: Getting Results*, 2e. Washington, DC: The Urban Institute Press.

Heavisides, B. and Price, I. (2001). Input versus output-based performance measurement in the NHS–the current situation. *Facilities* 19 (10): 344–356.

Herbert, R. and Jon, S. (2009). Service delivery and performance monitoring in PFI/PPP projects. *Construction Management and Economics* 27 (2): 181–197 (17).

Hibbard, J.H., Jewett, J.J., Legnini, M.W., and Tusler, M. (1997). Choosing a health plan: do large employers use the data? *Health Affairs* 16 (6): 172–180.

HM Treasury (2012). *UK Private Finance Initiative Projects: summary data as at March 2012 summary data as at March 2012*. https://www.gov.uk/government/uploads/system/ uploads/attachment_data/file/207369/summary_document_pfi_data_march_2012.pdf (accessed 18 November 2013).

Hope, J. and Fraser, R. (2003). *Beyond Budgeting: How Managers Can Break Free from the Annual Performance Trap*. Boston, MA: Harvard Business School Press.

Hurst, J. and Jee-Hughes, M. (2001). Performance measurement and PM in OECD Health Systems. OECD Labour Market and Social Policy Occasional Papers, No. 47, OECD Publishing. doi: 10.1787/788224073713 (accessed 3 November 2013)

Institute of Medicine (2001). *Committee on Quality of Health Care in America. Crossing the Quality Chasm: A New Health System for the 21st Century*. Washington, DC: National Academies Press.

International Facility Management Association (2003). What is facility management? http:// www.ifma.org/about/what-is-facility-management (accessed 27 November 2012).

Kaplan, R.S. and Norton, D.P. (1996). *The Balanced Scorecard: Translating Strategy into Action*. Boston, MA: Harvard Business School Press.

Kennerley, M. and Neely, A. (2003). Measuring performance in a changing business environment. *International Journal of Operations & Production Management* 23 (2): 213–229.

Kincaid, D.G. (1994). Measuring performance in facility management. *Facilities* 12 (6): 17–20.

King's Fund (2010). *Responding to the financial challenge facing the NHS*. http://www. kingsfund.org.uk/projects/general-election-2010/priorities/financial-challenges (accessed 25 August 2013).

King's Fund (2013). Patient-centred leadership: rediscovering our purpose. http://www. kingsfund.org.uk/sites/files/kf/field/field_publication_file/patient-centred-leadership-rediscovering-our-purpose-may13.pdf (accessed 13 November 2013).

Leahy, P. (2005). *Lessons from the private finance initiative in the United Kingdom*. EIB Papers, ISSN 0257-7755, 10 (2): 59–71.

Leatherman, S. and Sutherland, K. (2008). The quest for quality: refining the NHS reforms. Nuffield Trust. http://www.nuffieldtrust.org.uk/media-centre/press-releases/quest-quality-refining-nhs-reforms (accessed 16 October 2013).

Lennerts, K., Abel, J., Pfründer, U., and Sharma, V. (2003). Reducing health care costs through optimised facility management-related processes. *Journal of Facilities Management* 2 (2): 192–206.

Lichiello, P. and Turnock, B.J. (1999). *Guidebook for Performance Measurement*. New Jersey: Toining Point Publishing.

Loeb, J.M. (2004). The current state of performance measurement in health. *International Journal for Quality in Health Care* 16 (Supplement 1): 15–19.

Longenecker, C.O. and Fink, L.S. (2001). Improving management performance in rapidly changing organizations. *Journal of Management Development* 20 (1): 7–18.

Marshall, M., Wray, L., Epstein, P., and Grifel, S. (1999). 21st century community focus: better results by linking citizens, government and performance measurement. *Public Management* 81 (10): 12–19.

Martinez, J. (2000). Assessing Quality, Outcome and PM. World Health Organization, Department of Organization of Health Services Delivery Geneva Switzerland 2001.

Maxwell, J., Briscoe, F., Davidson, S., et al. (1998). Managed competition in practice: value purchasing by fourteen employers. *Health Affairs* 17 (3): 216–226.

McDowall, E. (2000). Monitoring PFI contracts. Facilities Management (December), pp. 8–9.

McGinnis, J.M., Williams-Russo, P., and Knickman, J.R. (2002). The case for more active policy attention to health promotion. *Health Affairs* (Millwood) 21 (2): 78–93.

McIvor, R., Wall, A., Humphreys, P., and McKittrick, A. (2009). *A Study of Performance Measurement in the Outsourcing Decision*. Oxford: CIMA Publishing.

Mecca, A.M. (1998). Monitoring Outcomes: Our New and Permanent Challenge. Centre for Substance Abuse Treatment TIE Communiqué.

Meng, X. and Minogue, M. (2011). Performance measurement models in facility management: a comparative study. *Facilities* 29 (11/12): 472–484.

Nani, A.J., Dixon, J.R., and Vollmann, T.E. (1990). Strategic control and performance measurement. *Journal of Cost Management* (Summer): 33–42.

NAO (National Audit Office) (2005). Darent Valley Hospital: The PFI Contract in Action, Report of Comptroller and Auditor General. HC 209, Session 2004–5, The Stationery Office, London.

Neath, A. (2010). *Measurement for quality and cost: Challenges, examples of success and working collectively*. NHS Institute for Innovation and Improvement. Available from: http://www.institute.nhs.uk/images//newsletter/2010-05-05%20WebEx_for_measurement_report.pdf (accessed 18 November 2013).

Neely, A. (1999). The performance measurement revolution: why now and what next? *International Journal of Operations & Production Management* 19 (2): 205–228.

Neely, A., Gregory, M., and Platts, K. (1995). Performance measurement system design: a literature review and research agenda. *International Journal of Operations & Productions Management* 15 (4): 80–116.

Nelson, R.R. and Winter, S.G. (1982). *An Evolutionary Theory of Economic Change*. Cambridge, MA: The Belknap Press.

Nerenz, D.R. and Neil, N. (2001). Performance measures for health care systems. Commissioned Paper for the Centre for Health Management Research May 1, 2001. *New Directions for Evaluation* 75 (Fall 1997): 5–13.

Newcomer, K.E. (1997). Using performance measurement to improve programs. *New Directions for Evaluation* 75: 5–13.

Ng, S.T. and Wong, Y.M.W. (2007). Payment and audit mechanisms for non-private-funded PPP-based infrastructure maintenance projects. *Construction Management and Economics* 25: 915–924.

NHS Institute for Innovation and Improvement (2008). http://www.institute.nhs.uk/quality_and_service_improvement_tools/quality_and_service_improvement_tools/balanced_scorecard.html#B (Assesse 8 November 2013).

Nies, H., Leichsenring, K., Veen, R., et al. (2010). *Quality management and quality. Assurance in long-term care European overview paper*. http://interlinks.euro.centre.org/sites/default/files/WP4_Overview_FINAL_04_11.pdf (accessed 18 October 2013).

OECD (2009). Government at a Glance. OECD Publishing. http://www.oecd.org/inclusive-growth/Government%20at%20a%20Glance%202011.pdf (accessed 12 October 2013).

O'Leary, D.S. (1995). Joint Commission on Accreditation of Healthcare Organizations. Measurement and accountability: taking careful aim. *Journal of Quality Improvement* 21 (July 1995): 354–357. Overview.

Oregon State University Family Study Centre (Clara C. Pratt, et al.) (1997). *Building Results: From Wellness Goals to Positive Outcomes for Oregon's Children, Youth, and Families*, 2e. Salem, OR: Oregon Commission on Children and Families.

Otley, D. (2002). In: *Business Performance Measurement: Theory and Practice* (ed. A.D. Neely). UK: Cambridge University Press.

Øvretveit, J. (2003). *What are the best strategies for ensuring quality in hospitals?* World Health Organization (WHO). http://citeseerx.ist.psu.edu/viewdoc/download?doi=10.1.1.126.8717 &rep=rep1&type=pdf (accessed 18 November 2013).

Parmenter, D. (2010). *Key Performance Indicators: Developing, Implementing, and Using Winning KPIs*, 2e. Hoboken, NJ: John Wiley & Sons, Inc.

Partnerships UK (2006). Report on Operational PFI Projects. March 2006. http://www.observatory.gr/files/meletes/Operational_PFI_PUK_report20060322.pdf (accessed 14 September 2013).

Pavlov, A. and Bourne, M. (2011). Explaining the effects of performance measurement on performance: an organizational routines perspective. *International Journal of Operations & Production Management* 31 (1): 101–122.

Perrin, E.B.J., Durch, S., and Skillman, S.M. (1999). *Health Performance Measurement in the Public Sector: Principles and Policies for Implementing an Information Network*. Washington, DC: National Academy Press.

Peters, T., and Waterman, R.H. (1982). *In Search of Excellence: Lessons from America's Best Run Companies*. UK: Bloomsbury.

Porter, M.E. (2008). What is value in health care? *The New England Journal of Medicine* 363: 2477–2481.

Pullen, S., Atkinson, D., and Tucker, S. (2000), Improvements in benchmarking the asset management of medical facilities. *Proceedings of the International Symposium on Facilities Management and Maintenance*, Brisbane, 265–271.

Purbey, S., Mukherjee, K., and Bhar, C. (2007). Performance measurement system for healthcare processes. *International Journal of Productivity and PM* 56 (3): 241–251.

Robinson, H.S. and Scott, J. (2009). Service delivery and performance monitoring in PFI/PPP projects. *Construction Management and Economics* 27 (2): 181–197.

Sachs, I.S. (2011). *Performance Driven IT Management: Five Practical Steps to Business Success*. Plymouth, UK: Government Institute Press.

Schneier, C.C., Shaw, D.G., Beatty, R.W. et al. (1995). *Performance Measurement, Management, and Appraisal Sourcebook*. MA: Human Resource Development Press, Inc.

Shohet, I.M. and Lavy, S. (2004). Healthcare facilities management: state of the art review. *Facilities* 22 (7/8): 210–220.

Silver, N. (2012). *The Signal and the Noise: Why So Many Predictions Fail–but Some Don't*. New York: Penguin Press.

Simons, R. (2000). *Performance Measurement and Control Systems for Implementing Strategy Text and Cases*. London, UK: Prentice–Hall International.

Slater, S.F., Olson, E.M., and Reddy, V.K. (1997). Strategy based performance measurement. *Business Horizons* 40 (Pt. 4): 37–44.

Spitzer, D.R. (2007). *Transforming Performance Measurement: Rethinking the Way We Measure and Drive Organisational Success*. New York, NY: AMACOM.

The Department of Health (NHS England) (2013). Patient-led assessment of the care environment (PLACE), high quality care for all, now and for future generation. http://www.england.nhs.uk/2013/02/19/place (accessed 17 October 2013).

The Health Foundation (2009). Quality Improvement Made Simple: What Every Board Should Know about Healthcare Quality Improvement. London: Health Foundation. http://www.health.org.uk/public/cms/75/76/313/594/Quality_improvement_made_simple.pdf?realName=uDCzzh.pdf (accessed 29 August 2013).

The Health Foundation (2013). *Is quality of care in England getting better? Quality Watch annual statement 2013: summary of findings*. www.qualitywatch.org.uk (Assesses 3 November 2013).

The HM Treasury (2003). *PFI: meeting the investment challenge*. www.hm-treasury.gov.uk (accessed 18 November 2013).

The National Health Service (NHS) (2004). *A matron's Charter: an action plan for cleaner hospitals*. http://www.rdehospital.nhs.uk/docs/patients/services/housekeeping_services/Matrons%20charter%20doc.pdf (accessed 17 August 2013).

Trisolini, M., Pope, G., Kautter, J., and Aggarwal, J. (2006) Medicare physician group practices: innovations in quality and efficiency. The Commonwealth Fund (December 2006).

Tucker, M. and Pitt, M. (2009). Customer performance measurement in facilities management: a strategic approach. *International Journal of Productivity and PM* 58 (5): 407–422.

Turner, R. (2008). *Handbook of Project Management*. 4e. Aldershot, UK: Gower.

U.S. Department of Energy programs (TRADE) (1995). *How to measure performance: a handbook of techniques and tools*. Oak Ridge Associated Universities, USA.

U.S. Department of Health and Human Services (2006). *Substance abuse and mental health services administration centre for substance abuse treatment*. http://www.ncbi.nlm.nih.gov/books/NBK64075/pdf/TOC.pdf (accessed 13 October 2013).

Varcoe, B.J. (1993). Facilities performance: achieving value for money through performance measurement and benchmarking. *Property Management* 11 (4): 301–307.

Varcoe, B.J. (1998). Facilities as a production asset. Facilities Management (April), pp.6–8.

Ventovuori, T. (2006). Elements of sourcing strategies in FM services-a multiple case study. *International Journal of Strategic Property Management* 10: 249–267.

Wang, Q. (1992). Activity-Based Facility Management Benchmarking, private communication.

Welch, S. and Mann, R. (2001). The development of a benchmarking and performance improvement resource. *Benchmarking: An International Journal* 8 (5): 431–452.

WHO (2006). *Quality of care: a process for making strategic choices in health systems?* http://www.who.int/management/quality/assurance/QualityCare_B.Def.pdf (accessed 18 November 2013).

Wilcock, P.M. and Thomson, R.G. (2000). Modern measurement for a modern health service. *Quality in Health Care* 9: 199–202.

## Further Reading

Atkinson, H. and Brown, J.B. (2001). Rethinking performance measures: assessing progress in UK hotels. *International Journal of Contemporary Hospitality Management* 13 (3): 128–135.

Binder, C. (2001). Measurement: a few important ideas. *Performance Improvement* 40 (3): 20–28.

Brown, M.G. (1994). Is your measurement system well-balanced? *Journal for Quality and Participation* 17: 6–11.

Brown, M.G. (1996). *Keeping Score: Using the Right Metrics to Drive World-Class Performance*. New York, NY: Quality Resources.

Cable, J.H. and Davis, J.S. (2005), *Key performance indicators for federal facilities portfolios*: Federal Facilities Council technical report number 147 in conjunction with the Federal Facilities Council Ad Hoc. Committee on Performance Indicators for Federal Real Property Asset Management, National Research Council.

Cain, C.T. (2004). *Performance Measurement for Construction Profitability*. Oxford: Blackwell Publishing Ltd.

Chan, F.T.S. (2003). Performance measurement in a supply chain. *The International Journal of Advanced Manufacturing Technology* 21: 534–548.

Crosby, P.B. (1984). *Quality without Tears*. New York, NY: McGraw-Hill. Cited in health care financing in McIntyre, D., Rogers, L., & Heier, E. J., (2001). Overview, History, and Objectives of Performance Measurement review Volume 22, Number 3.

David, C. (2006). The United Kingdom private finance initiative: the challenge of allocating risk. OECD Journal on Budgeting 5 (3). http://www.oecd.org/unitedkingdom/43479923. pdf (accessed 18 November 2013).

Dennis, S.O. (1995). Joint Commission on Accreditation of Healthcare Organizations. Measurement and accountability: taking careful aim. *Journal of Quality Improvement* 21 (July): 354–357.

Department of Treasury and Finance: Government of Western Australia (2002). *Partnership for growth: policies and guidelines for Public Private Partnership in Western Australia*. www. dtf.wa.gov.au (accessed 15 July 2007).

Douglas, J. (1996). Building performance and its relevance to facilities management. *Facilities* 14 (3/4): 23–32.

Eagle, C.J. and Davies, J.M. (1993). Current models of 'quality' – an introduction for anaesthetists. *Canadian Journal of Anaesthesia* 40: 851–862.

Eddy, D.M. (1998). Performance measurement: problems and solutions. *Health Affairs* (Millwood) 1998; 17 (4): 7–25.

Edet, C. and Gidado, K. (2004). Evaluating the operation of PFI in roads and hospitals. Association of Chartered Certified Accountants, Research report No.84.

Edet, C.B. and Gidado, K. (2008). PFI hospitals in the UK: measuring the medical practitioners' level of satisfaction. In: *Proceedings 24th Annual ARCOM Conference* (ed. A. Dainty), (1–3 September 2008). Cardiff, UK: Association of Researchers in Construction Management, 527–534.

Elg, M., Broryd, K.P., and Kollberg, B. (2013). Performance measurement to drive improvements in healthcare practice. *International Journal of Operations & Production Management* 33 (11/12): 1623–1651.

Enoma, A. and Allen, S. (2007). Developing key performance indicators for airport safety and security. *Facilities* 25 (7): 296–315.

ERIC (2010). https://www.gov.uk/government/publications/hospital-estates-and-facilities-statistics-2010-11 (accessed 29 August 2013).

Federal Trade commission and the Department of Justice (2004). *Improving healthcare: a dose of competition.* http://www.ftc.gov/reports/healthcare/040723healthcarerpt.pdf (accessed 18 November 2013).

Feurer, R. and Chaharbaghi, K. (1995). Performance measurement in strategic change. *Benchmarking for Quality Management & Technology* 2 (2): 68–83.

Fitzgerald, L., Johnston, R., Bignall, S., et al. (1993). *Performance Measurement in Service Businesses.* Chartered Institute of Management Accountants.

Francis, R. (2013a). *Report of the Mid Staffordshire NHS Foundation Trust Public Inquiry.* Department of Health. https://francisresponse.dh.gov.uk (accessed 20 October 2013).

Francis, R. (2013b). *Report of the Mid Staffordshire NHS Foundation Trust Public Inquiry.* The Stationery Office.

Froud, J. and Shaoul, J. (2001). Appraising and evaluating PFI for NHS hospitals. *Financial Accountability and Management* 17 (3): 247–270.

Gaffney, D. and Pollock, A. (1999). Pump priming the PFI: why are privately financed hospital schemes being sub- sidized. *Public Money & Management* 17 (3): 11–16.

Ginsberg, C. and Sheridan, S. (2001). Limitations of and barriers to using performance measurement: purchasers' perspectives. *Health Care Financing Review* 22 (3).

Green, A. (1998). Model facilities. *Facilities Management* (April), pp. 12–13.

Grout, P.A. (1997). Economics of the private finance initiative. *Oxford Review of Economic Policy* 13 (4): 53–66.

H.M. Treasury (2004). *Value for Money Assessment Guidance.* London: HMSO.

H.M. Treasury (2006). *Value for Money Assessment Guidance.* London: HMSO.

Halachmi, A. (2005). Performance measurement is only one way of managing performance. *International Journal of Productivity and PM* 54 (7): 502–516.

Hall, S. (2013). *Private finance initiative: PFI projects cost £2.4bn.* BBC South West. http://www.bbc.co.uk/news/uk-england-devon-22355993 (accessed 18 October 2013).

Harrowell, J. (2011). *Aged care standards and accreditation agency.* Response to the productivity commission draft report: caring for older Australians. http://www.pc.gov.au/__data/assets/pdf_file/0006/108555/subdr763.pdf (accessed 12 October 2013).

Harry, P.H. (1997). Where the rubber meets the road: performance measurement for state and local public agencies. *New Directions for Evaluation* 75: 31–44.

Hazell, M. and Morrow, M. (1992). Performance measurement and benchmarking. Management Accounting (December 9), pp. 44–45.

Hinks, J. and McNay, P. (1999). The creation of a management-by-variance tool for facilities management performance assessment. *Facilities* 17 (1/2): 31–53.

Hudson, M., Smart, A., and Bourne, M. (2001). Theory and practice in SME performance measurement systems. *International Journal of Operations & Productions Management* 21 (8): 1096–1115.

Ibrahim, J.E. (2001). Performance indicators from all perspectives. *International Journal of Quality in Health Care* 13: 431–432.

Igal, M. and Shohet, S.L. (2004). Healthcare facilities management: state of the art review. *Facilities* 22 (7/8): 210–220.

Journal of Institute of Healthcare Engineering and Estate Management. http://www. healthestatejournal.com/Print.aspx?Story=7471 (accessed 1 November 2013).

Judson, A.S. (1990). *Making Strategy Happen, Transforming Plans into Reality*. London: Basil Blackwell.

Kadefors, A. (2008). Contracting in FM: collaboration, coordination and control. *Journal of Facilities Management* 6 (3): 178–188.

Kagioglou, M., Cooper, R., and Aouad, G. (2001). PM in construction: a conceptual framework. *Construction Management & Economics* 19: 85–95.

Kaplan, R.S. and Norton, D.P. (1992). The balanced scorecard: measures that drive performance. Harvard Business Review (January/February), pp. 71–79.

Keegan, D.P., Eiler, R.G., and Jones, C.P. (1989). Are your performance measures obsolete? Management Accounting (June), pp. 45–50.

Kennerley, M. and Neely, A. (2001). Enterprise resource planning: analysing the impact. *Integrated Manufacturing Systems* 21 (2): 103–118.

Keogh, B. (2013). *Review into the quality of care and treatment provided by 14 hospital trusts in England: overview report*. http://www.nhs.uk/NHSEngland/bruce-keogh-review/ Documents/outcomes/keogh-review-final-report.pdf (accessed 17 October 2013).

Keser, C. and Willinger, M. (2007). Theories of behaviour in principal-agent relationships with hidden action. *European Economic Review* 51: 1514–1533.

Laitinen, E.K. (2002). A dynamic performance measurement system: evidence from small Finnish technology companies. *Scandinavian Journal of Management* 18: 65–99.

Letza, S.R. (1996). The design and implementation of the balanced business scorecard: an analysis of three companies in practice. *Business Process Re-engineering & Management Journal* 2 (3): 54–76.

Lim, E.C. and Alum, J. (1995). Construction productivity: issues encountered by contractors in Singapore. *International Journal of Project Management* 13 (1): 51–68.

Lingle, J.H. and Schiemann, W.A. (1996). From balanced scorecard to strategy gauge: is measurement worth it? Management Review (March), pp. 56–62.

Lohr, K. (1990). *Medicare: A Strategy for Quality Assurance* (ed.). Washington, DC: National Academy Press.

Lynch, R. and Cross, K. (1991). *Measure Up! Yardsticks for Continuous Improvement*. Oxford: Blackwell.

Maisel, L.S. (1992). Performance measurement: the balanced scorecard approach. *Journal of Cost Management* 5 (2): 47–52.

Martin, L.A., Nelson, E.C., Lloyd, R.C., and Nolan, T.W. (2007). Whole System Measures. IHI Innovation Series White Paper. Cambridge, MA: Institute for Healthcare Improvement. www.IHI.org.

Martin, L.L. 2004. Bridging the gap between contract service delivery and public financial management: applying theory to practice. In: *Financial Management Theory in the Public Sector* (eds. A. Khan and W.B. Hildreth), 55–70. Portsmouth, NH: Greenwood Publishing Group.

Martinez, V. (2005). Performance Measurement Systems: Mix Effects. http://euram2005. wi.tum.de (accessed 18 May 2011).

Massheder, K. and Finch, E. (1998a). Benchmarking methodologies applied to UK facilities. *Management Facilities* 16 (3–4): 99–106.

McDougall, G. and Hinks, J. (2000). Identifying priority issues in facilities management benchmarking. *Facilities* 18 (10/11/12): 427–434.

McDowall, E. (1999). Specifying performance for PFI. Facilities Management (June), pp. 10–11.

McGlynn, E.A., Asch, S.M., Adams, J., et al. (2003). The quality of health care delivered to adults in the United States. *The New England Journal of Medicine* 348 (26): 2635–2645.

Mckevith, D. and Lawton, A. (2001). *Public Sector Management: Theory, Critique & Practice.* London, UK: SAGE Publication.

Medori, D. and Steeple, D. (2000). A framework for auditing and enhancing performance measurement systems. *International Journal of Operations & Production Management* 20 (5): 520–533.

Meekings, A. (1995). Unlocking the potential of performance measurement: a practical implementation guide. *Public Money and Management* 15 (4): 5–12.

Meyer, C.B. (2001). A case in case study methodology. *Field Methods* 13 (4): 329–352.

Miles, M.B. (1994). *Qualitative Data Analysis: An Expanded Sourcebook*, 2e. Thousand Oaks, CA: Sage.

Miller, T. and Leatherman, S. (1999). The National Quality Forum: a me-too or a breakthrough in quality measurement and reporting? *Health Affairs* 8 (6): 233–237.

Nanni, A.J., Dixon, J.R., and Vollmann, T.E. (1992). Integrated performance measurement: management accounting to support the new manufacturing realities. *Journal of Management Accounting Research* 4: 1–19.

National Advisory Group on the Safety of Patients in England (2013). *A promise to learn, and a commitment to act: improving the safety of patients in England.* https://www.gov.uk/government/uploads/system/uploads/attachment_data/file/226703/Berwick_Report.pdf (accessed 18 October 2013).

National Health Performance Committee (2004). National Report on Health Sector Performance Indicators. http://www.aihw.gov.au/publication-detail/?id=6442467669 (accessed 18 November 2013).

Neely, A. (1998). *Measuring Business Performance.* London: Economist Books.

Neely, A. (2002). *Business Performance Measurement, Theory and Practice.* Cambridge: Cambridge University Press.

Neely, A. (2005). The evolution of performance measurement research: developments in the last decade and a research agenda for the next. *International Journal of Operations & Production Management* 25 (12): 1264.

Neely, A., Adams, C., and Crowe, P. (2001). The performance prism in practice measuring excellence. *The Journal of Business PM* 5 (2): 6–12.

Neely, A., Mills, J.F., Gregory, M.J. et al. (1996). *Getting the Measure of Your Business.* London, UK: Findlay.

NHS National Survey (2013). http://www.nhssurveys.org (accessed 18 November 2013).

Nightingale, F. (1992). In: *Notes on Nursing* (ed. V. Skretkowitz). London: Scutari Press.

Norreklit, H. (2000). The balance on the balanced scorecard – a critical analysis of some of its assumptions. *Management Accounting Research* 11: 65–88.

Norreklit, H. (2003). The balanced scorecard: what is the score? A rhetorical analysis of the balanced scorecard. *Accounting, Organizations and Society* 28 (6): 591–619.

Otley, D. (1999). PM: a framework for management control systems research. *Management Accounting Research* 10: 363–382.

Otley, D. (2001). Extending the boundaries of management accounting research: developing systems for PM. *The British Accounting Review* 33 (3): 243–261.

Parida, A. and Kumar, U. (2006). Maintenance Performance Measurement (MPM): issues and challenges. *Journal of Quality in Maintenance* 12 (3): 239–251.

Parker, C. (2000). Performance measurement. *Work Study* 49 (2): 63–66.

Patel, M. and Robinson, H. (2010). Impact of governance on project delivery of complex NHS PFI/PPP schemes. *Journal of Financial Management of Property and Construction* 15 (3): 216–234.

Patton, M.Q. (1990). *Qualitative Evaluation and Research Methods*, 2e. Thousand Oaks, CA: Sage Publications, Inc.

Pollock, A., Shaoul, J., and Vickers, N. (2002). Private finance and value for money in NHS hospitals: a policy in search of a rationale? *British Medical Journal* 324: 1205–1209.

Porter, M.E. (2007). Defining and introducing value in health care. *Evidence-Based Medicine and the Changing Nature of Health Care: IOM Annual Meeting* Summary (2008), 161–172.

Porter, M.E. and Molander, R. (2010). *Outcomes measurement: learning from international experiences*. Institute for Strategy and Competitiveness, Harvard University Working Paper. http://www.hbs.edu/rhc/prior.html (accessed 9 October 2013).

Preiser, W.F.E. (1995). Post occupancy evaluation: how to make buildings work better. *Journal of Facilities* 13 (11): 19–28.

Pryke, S.D. and Smyth, H.J. (2006). *Management of Complex Projects: Relationship Approach*. Oxford: Blackwell.

Rouse, P. and Putterill, M. (2003). An integral framework for performance measurement. *Management Decision* 41 (8): 791–805.

Shapiro, J. and Shapiro, R. (2003). Towards an improved collaboration model for the national healthcare system in England and Wales: a critical and constructive approach using operational research. *Logistics Information Management* 16 (3/4): 246–258.

Shohet, I.M. (2003a). Building evaluation methodology for setting priorities in hospital buildings. *Construction Management and Economics* 21 (7): 681–692.

Shohet, I.M. (2003b). Key performance indicators for maintenance of health-care facilities. *Facilities* 21 (1/2): 5–12.

Skretkowitz, V.E. (1992). *Nightingale, F. Notes on Nursing*. London: Scutari Press.

Slack, N. (1991). *The Manufacturing Advantage, Mercury Books*. London. WHO (2000), WHR2000. Geneva: World Health Organization.

Smith, P. (1995). In: *Performance Measurement and Evaluation* (ed. J. Holloway, J. Lewis, and G. Mallory), 192–216. London: Sage, Chapter 10.

Smyth, H. and Edkins, A. (2007). Relationship management in the management of PFI/PPP projects in the UK. *International Journal of Project Management* 25: 232–240.

Tangen, S. (2003). An overview of frequently used performance measures. *Work Study* 52 (7): 347–354.

The Drucker Institute, Peter Drucker's Life and Legacy. Available from Peter Drucker Institute. http://www.druckerinstitute.com/link/about-peter-drucker (accessed 27 August 2013.

The Economist (2013). Doing more with less. http://graphics.eiu.com/upload/eb/BMI_Doing_more_with_less_WEB.pdf (accessed 22 August 2013).

The HM Treasury (2000). PPPs: The Government's Approach (2000). www.hm-treasury.gov.uk (accessed 12 October 2013).

The National Performance Review (1997). *Servig the American Public: Best Practices in Performance Measurement, Benchmarking Study Report*. Washington, DC: Government Printing Office.

The National Performance Review (now the National Partnership for Reinventing Government) (1997). *Serving the American Public: Best Practices in Performance Measurement. Benchmarking Study Report*. Washington, DC: Government Printing Office.

Thompson, B.L. and Harris, J.R. (2001). Performance measures: are we measuring what matters? *American Journal of Preventive Medicine* 20: 291–293.

Thomson, R.G. and Lally, J. (2000). PM at the crossroads in the NHS: don't go into the red. *Quality in Health Care* 2000; 9: 199–202.

Turnock, B. and Handler, A. (1997). From measuring to improving public health practice. *Annual Review of Public Health* 1997; 18: 261–282.

Tversky, A. and Daniel, K. (1986). Rational choice and the framing of decisions. *Journal of Business* 59 (4), Part 2: 5251–5278.

Ulrich, B.T. (1992). *Leadership and Management According to Florence Nightingale*. Norwalk, CT: Appleton & Lange.

U.S. Department of Health and Human Services (2006). *Substance abuse and mental health services administration centre for substance abuse treatment*. http://www.ncbi.nlm.nih.gov/books/NBK64075/pdf/TOC.pdf (accessed 13 October 2013).

U.S. Department of Justice (2004). Improving healthcare: a dose for competition. http://www.ftc.gov/reports/healthcare/040723healthcarerpt.pdf. (accessed 12 November 2012).

Vliet, A. (1997). Are they being served? Management Today (February), pp.66–69.

Voss, C., Tsikriktsis, N., and Frohlich, M. (2002). Case research in operations management. *International Journal of Operations & Production Management* 22 (2): 195.

Wang, L., Yu, C.W., and Wen, F.S. (2007). Economic theory and the application of incentive contracts to procure operating reserves. *Electrical Power Systems Research* 77: 518–526.

Wholey, J.S. and Hatry, H.P. (1992). The case for performance monitoring. *Public Administration Review* 52: 604–610.

Wilson, D. (2013). *The problem with benchmarking*. http://www.fm-world.co.uk/good-practice-legal/legal-articles/the-problem-with-benchmarking. (accessed 18 November 2013).

Winch, G. (2000). Institutional reform in British construction: partnering and private finance. *Building Research & Information* 28 (2): 141–155.

Wisner, J.D. and Fawcett, S.E. (1991). Link firm strategy to operating decisions through performance measurement. *Production and Operations Management Journal* 32 (3): 5–11.

Witcher, B.J. and Chau, V.S. (2007). Balanced scorecard: dynamic capabilities for managing strategic fit. University of East Anglia UK, *Management Decision* 45 (3): 518–538.

# 5

# Procurement and Contracting for Maintenance and Refurbishment Works

## 5.1 Introduction

This chapter looks at the processes associated with the procurement and contracting of refurbishment and maintenance works. The latest BCIS market analysis (BCIS 2021) suggests the market to be worth £66bn. It is therefore of critical importance that the way we buy from the construction industry ultimately decides the success or failure of the project. A report by Glenigan (2015) found only 69% of projects were completed to, or lower than, the budget and 40% of projects achieved time targets. Yet worse still is discovered when client satisfaction with the outcome of the project is reviewed with alarming levels of dissatisfaction and reports of poor project performance. This chapter introduces strategic procurement as a tool to elicit project performance requirements, making the case for a businesslike approach to procurement. The chapter then proceeds to introduce the various procurement routes adopted for refurbishment and maintenance projects before providing an overview of procurement governance requirements as the project is let within the marketplace. The second section of the chapter deals with contracts and contract selection. National Building Specification (NBS 2018) conducted the National Construction Contracts and Law Survey of employers, consultants and contractors over the period August to November 2017. Based upon 360 responses, it reported that the most popular contract between employers and contractors is the Joint Contract Tribunal (JCT) suite, accounting for 62% of contracts used by their respective organisations. The second most popular was the New Engineering Contract (NEC), accounting for 14% of contracts. Other forms of contract including bespoke, FIDIC, PPC2000 and others made up the remaining 24%. Given the dominance of the JCT and NEC in the UK construction market, this chapter will conclude by introducing the JCT and NEC contract suites before exploring the various contract options with each suite suitable for the management of building maintenance and refurbishment projects.

## 5.2 Rationale for Procurement of Maintenance Interventions

Before a discussion of procurement can take place, it is firstly important to understand the drivers for maintenance and the nature of building maintenance. Although this is outlined in other chapters of this book, it is nevertheless helpful to revisit these issues

*Introduction to Built Asset Management*, First Edition. Dr Anthony Higham, Dr Jason Challender, and Dr Greg Watts.
© 2022 John Wiley & Sons Ltd. Published 2022 by John Wiley & Sons Ltd.

from the perspective of procurement. The first key issue is to understand what is being procured. The scope of this book covers a range of possibilities that include either one or several of the items listed below:

- **Repairs** – Defined by the British Standards Institution as 'actions necessary to return a building or its parts to an acceptable condition by the renewal, replacement or mending of worn, damaged or degraded parts' (British Standards Institution 2000, p. 5).
- **Maintenance** – Maintenance can be defined as the actions needed to carry out repairs; however, ISO 6707-1 suggests maintenance to be the 'combination of all technical and associated administrative actions during the service life to retain a building or its parts in a state in which it can perform its required functions' (British Standards Institution 2000, p. 5).
- **Refurbishment and rehabilitation** – Mansfield (2008) considers refurbishment to be a larger form of maintenance intervention triggered by one of four situations:
  - ○ The incidence of profound damage to the physical structure.
  - ○ In anticipation of deterioration in a major element (triggered by the condition survey).
  - ○ Undertaken in anticipation of leading a changing market (demand led).
  - ○ Undertaken to secure a position in a recently changed property market.

  On the other hand, Mansfield also purports that refurbishment is implemented to extend the beneficial use of an existing building by providing a cost-effective alternative to redevelopment, thus providing a positive counter to the negative processes of physical deterioration and obsolescence that are the principle causes of depreciation (the loss of a properties investment value).

The nature of the intervention or indeed the series of interventions to be procured will have a significant impact on the approach adopted to the procurement process. It is critically important for the team to distinguish between maintenance/repair activities and refurbishment/rehabilitation. The traditional view of funding within organisational budget structures has always been that maintenance activities are drawn from the *revenue account* and refurbishment or other associated activities will be drawn from the *capital account*. Nevertheless, the procurement process will involve key strategic activities with property and construction professionals needing to furnish their organisation's senior leadership with a strong business case for all maintenance or refurbishment interventions. For more detail about the business case please refer to Chapter 6, Section 6.2.

## 5.3 The Procurement Process

Seeking to appoint contractors and manage the overall process of letting a building maintenance or refurbishment intervention is no different to any other form of procurement that takes place in the construction sector. However, it is often noted that little consideration is given to the strategic perspectives of procurement; instead, those tasked with implementing procurement often jump straight into procurement routes. It is, therefore, vitally important that procurement is understood as a business strategy, before the specifics of procurement routes and contract options are appraised. Starting this process from the very beginning, it is firstly important to define the concept of procurement. In the introduction it was mentioned that procurement is the way we 'buy from the construction industry', which it essentially is. However, more comprehensive definitions provide a better understanding of why this process needs to be driven strategically and from a business perspective. The international standard for Construction Procurement, BS ISO 10845-1 (British Standards Institution 2020), defines construction procurement as 'the process through which contracts are created, managed and fulfilled', before going on to suggest there are six principal activities associated with the procurement process, including:

- Establishing what is to be procured
- Deciding on procurement strategies in terms of packaging, contracting and targeting strategy and selection method
- Soliciting tender offers
- Evaluating tender offers
- Awarding contracts
- Administering contracts and confirming compliance with requirements

Again, as outlined in the introduction, getting this process right will ultimately decide the success or failure of the project/maintenance works being procured. Regrettably, the construction sector's track record in successful project delivery is somewhat pitiful. As mentioned earlier, a report by Glenigan (2015) found that only 69% of projects were completed to, or lower than, the budget and 40% of projects achieved time targets. It is, therefore, of critical importance that procurement is approached as a strategic process, with the development of a procurement strategy or policy that identifies the best way of achieving the organisation's requirements. BS8534:2011 describes a procurement policy as a working document that 'details the overarching strategic principles and objectives for procurement of construction projects; spanning the whole project life cycle from identification of needs through to the end of the useful life of an asset' (British Standards Institution 2011, p. 3). A better way of understanding the procurement policy is to consider it in a visual format, as shown in Figure 5.1.

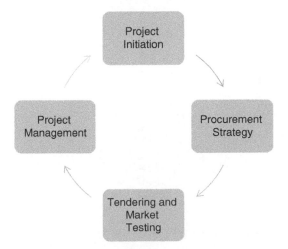

**Figure 5.1**   High-level procurement cycle. (Adapted from BSI 2011).

The figure illustrates four key stages to effective procurement; three of these stages provide the skeleton framework for the development of this chapter of the book. Whilst not every element of each stage will be explored in detail, key features of the procurement process will be identified and as the chapter progresses. The final, fourth stage of the procurement process, the management of maintenance interventions, is explored in detail in Chapter 10.

## 5.4 Project Initiation

At this stage in the procurement cycle, careful strategic planning should be undertaken to identify the key issues and delivery options in relation to procurement. With this in mind BS8534 identifies three key deliverables for this stage in the procurement process, as shown in Figure 5.2.

The first stage in procurement initiation is to establish the business need. This will include the establishment of why procurement activities are needed and the skills, capabilities and resources available to the organisation to bring about the effective procurement. For routine maintenance, the organisation will have considered issues such as the possibility of retaining the responsibility in house or the benefits of potential outsourcing. However, the need for maintenance, as outlined earlier in the book, cannot be disputed. For capital projects such as planning maintenance or some level

**Figure 5.2** Attributes of project initiation (Adapted from BSI 2011).

of refurbishment, the business need will often be appraised alongside the business case for proceeding with the procurement. A detailed discussion of the project business case is provided in Chapter 6 (Section 6.2), so it is not proposed to repeat that discussion here.

Key to successful procurement initiation and to an extent the development of a successful business case is the determination of project objectives and project success measures. There is a lot of academic debate about the notion of project success and how project success is determined, depending on the perspective of the stakeholder questioned. Yet for the most part, client organisations and professionals working within the construction sector continue to view project success from the project management perspective of time, cost and quality. As depicted in Figure 5.3, this perspective is often presented as a triangle, with decision makers often facing trade-off decisions between the three attributes.

This triangle appears simple, and in some respects it is. For instance, time is clearly linked to the duration of the project with performance measured against objectives typically outlined on a programme of works or via process gateways. Whilst cost is again a reasonably simple construct to define, as outlined in Chapter 6, the financial performance of the item to be procured can be appraised with budgets devised and monitored, and overruns in price ascertained and reported. The challenges begin when quality has to be defined. Quality as a concept has always been difficult to define. To their credit, employers are becoming much more aware of the need to accurately define and measure performance in this important area. However, for some, quality will continue to be narrowly defined and mapped against specification and drawings with quality robustly monitored during construction to ensure specification compliance and building safety. For some clients, a far wider measure of quality will be developed. Such

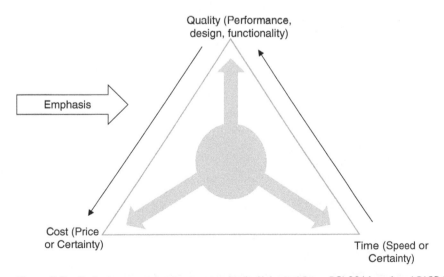

**Figure 5.3** Project management success criteria (Adapted from BSI 2011, p. 6 and RICS 2013, p. 4).

perspectives can include measures such as health and safety performance, environmental protection, customer or user satisfaction with repair response times and the politeness of the maintenance operative, or the benefits via social value attributes that the contractor brings to the local community. These are collectively captured and monitored through the use of key performance indicators that can be outlined at the earliest stage in the project inception and monitored through its delivery cycle. Whether this is a five-year reactive maintenance contract or a refurbishment project, the process is essentially the same.

## 5.5  Procurement Strategy

Once the need for the project has been established, the next fundamental stage in procurement is to define and develop a procurement policy or strategic framework. The strategy identifies the best way of achieving the completion of the project. Whilst completion is possible for larger capital projects, others such as reactive maintenance contracts will, of course, not have an end but will have completion in the sense of a contract duration, e.g., three years. Either way, the procurement strategy will make provision for careful strategic planning focused on identifying key issues and delivery options in relation to procurement, including the achievement of an optimum balance of risk, control and funding for a project. However, the choice of a particular procurement strategy largely depends on a client's required balance of cost, quality and time risks (RICS 2013). With this in mind, BS8534 identifies a series of key deliverables for this stage in the procurement process, as shown in Figure 5.4 and considered in more detail in the ensuing sections of this chapter.

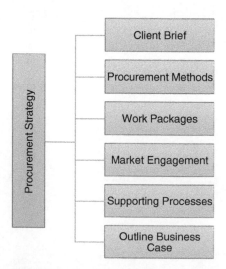

**Figure 5.4**   Key features of the procurement strategy (Adapted from BSI 2011).

## 5.6   Client Brief

The first stage in the development of the procurement strategy is to return to the client and establish their attitudes and key drivers. Some of these will feature in the business case. The procurement strategy will undoubtably include a series of objectives alongside important project context. Through this process important details relating to project scope, client priorities and project parameters will have emerged. Despite this, it remains essential that a more detailed client brief is established to allow the professional team to clarify the objectives, understand what the client is expecting in terms of outcomes for the project and how these are to be prioritised within the project management framework outlined in Figure 5.3. It is widely accepted that it is impossible to deliver against all three headline outcomes (cost, time and quality/performance), so trade-offs within the prioritisation process become necessary. As a result, important questions arise, such as how much the client wishes to spend, what level of performance, design or functionality are they seeking to achieve and where are the trade-offs to be made, for instance, the importance of enhanced functionality over design aesthetic. The brief should also address the client's risk appetite; this will include guidance on the client's ability to absorb risk or their intent to transfer risk and at what level this will need to be accommodated. Fundamentally, however, the brief should provide sufficient detail and guidance to allow the project team to undertake any design and specification works needed to advance the project to delivery and occupation. In this regard, BS8534 (BSI 2011) advises that the brief should at a minimum:

- Provide accurate statements reflecting the requirements of the project.
- Provide a basis from which the project can be initiated.
- Indicate how the acceptability of the finished product might be assessed.
- Take account of health and safety requirements.
- Provide details of any sustainability issues such as BREEAM performance levels (BREEAM is the enviromental assessment framework developed by the British Research Establishment as a means to quantify the sustainability credentials of a series of buildings), social value requirements, etc.
- Outline any design quality considerations or minimum requirements.

To fully ensure the brief meets the client's requirements, the RICS guidance note *Developing a Construction Procurement Strategy and Selecting an Appropriate Route* encourages the team to take their time with the client briefing process and build in opportunities for the client, or rather the client's senior team, to reflect on the issues emerging from the brief and to allow them time to ensure the emerging brief maps to the business needs for the project. Furthermore, the RICS (2013, p. 21) advocates the use of *play back* where the brief is translated into high-level construction outputs, such as:

- An example of a building at a stated cost with high levels of cost certainty.
- An example building of a certain size, shape, appearance and performance.
- An example of a building constructed within a specified timescale.

Each of these high-level examples could then be taken away by the client, allowing them time to reflect on how well each option would satisfy the business case both in terms of financial, time and other performance metrics, and how well each would satisfy the identified organisational needs. This would then allow the brief to be tweaked accordingly before the project moved forwards into design and development.

## 5.7 Procurement Route

Procurement routes or procurement systems are often the focus of many 'procurement modules' studied within built environment undergraduate or postgraduate programmes of study. When you think of the word 'procurement', it is the procurement route that typically leaps to mind. Yet in reality, the procurement route is only a small part of the overall procurement process. In their Achieving Excellence in Construction Procurement: Procurement and Contract Strategies guidance note, the Office of Government Commerce (2003) sought to clarify this issue by explaining the relationship between procurement strategy, procurement route and contract strategy. In this explanation it was suggested the procurement strategy determines the most appropriate procurement route, including contract strategy to fit the project objectives and current circumstances. The procurement route delivers the procurement strategy and will include an appropriate contract strategy that will best meet the client's needs in terms of risk allocation, delivery and financial performance.

### 5.7.1 Categorisation of Procurement Routes and Pricing Mechanisms

Despite being a common feature in literature, the categorisation of procurement routes is a difficult process to undertake. Categories can vary between authoritative sources, such as prominent authors and professional bodies. One key source is the Royal Institution of Chartered Surveyors (RICS) *Contracts in Use* survey, which provided a longitudinal perspective on procurement in the construction market from 1985 until its final publication in 2012. Since 2012, surveys of a similar type have been conducted by the Royal Institute of British Architects via NBS, with the latest survey published in 2018. The RICS procurement classifications together with the volume of use, and use by number in the UK in 1985, 2004 and 2010 are shown in Tables 5.1 and 5.2. Trends can be observed in the data, such as the fluctuating popularity of traditional procurement, based on a bill of quantities, alongside the seeming dominance of both traditional and design and build procurement routes.

Whilst the RICS classify lump sum, target cost, remeasurement and prime cost as procurement systems, the NBS calls some of these same approaches pricing mechanisms with only design and build and traditional identified as dominant procurement routes with trend data for these two main approaches identified in Table 5.3. As with the data from the RICS surveys, looking across the longitudinal data sample there is a slow move away from traditional procurement with 72% of consultants using it most often in 2012, dropping to 48% by 2018. Similar trends are reported by clients and contractors who both report a reduced appetite for traditional procurement methods. Interestingly, the data does not show a corresponding increase in design and build, suggesting some growth in the more niche approaches to procurement is likely; however, the NBS survey has not historically published data for these approaches so it is impossible to say whether or not this has been the case.

**Table 5.1** Trends in procurement – by number of contracts (RICS 2012).

| Procurement Method | 1985(%) | 2004(%) | 2007(%) | 2010(%) |
|---|---|---|---|---|
| Lump sum – Bill of Quantities | 42.8 | 31.1 | 20.0 | 24.5 |
| Lump sum – Specification and Drawing | 47.1 | 42.7 | 47.2 | 52.1 |
| Lump sum – Design and Build | 3.6 | 13.3 | 21.9 | 17.5 |
| Target Cost | 0 | 6.0 | 4.5 | 3.7 |
| Remeasurement – Approx. Bill of Quantities | 2.7 | 2.0 | 1.7 | 0.3 |
| Price Cost plus Fixed Fee | 2.1 | 0.2 | 0.5 | 0.6 |
| Management Contracting | 1.7 | 0.2 | 0.5 | 0 |
| Construction Management | 0 | 0.9 | 1.1 | 0.3 |
| Partnering Agreements | 0 | 2.7 | 2.4 | 1.0 |
| Total | 100 | 100 | 100 | 100 |

**Table 5.2** Trends in procurement – by value of contracts (RICS 2012).

| Procurement Method | 1985(%) | 2004(%) | 2007(%) | 2010(%) |
|---|---|---|---|---|
| Lump sum – Bill of Quantities | 59.3 | 23.6 | 13.2 | 18.8 |
| Lump sum – Specification and Drawing | 10.2 | 10.7 | 18.2 | 22.6 |
| Lump Sum – Design and Build | 8.0 | 43.2 | 32.6 | 39.2 |
| Target Cost | 0 | 11.6 | 7.0 | 17.1 |
| Remeasurement – Approx. Bill of Quantities | 5.4 | 2.5 | 2.0 | 0.7 |
| Price Cost plus Fixed Fee | 2.7 | 0.1 | 0.2 | 0.6 |
| Management Contracting | 14.4 | 0.8 | 1.0 | 0 |
| Construction Management | 0 | 0.9 | 9.6 | 0.1 |
| Partnering Agreements | 0 | 6.6 | 15.6 | 0.9 |
| Total | 100 | 100 | 100 | 100 |

**Table 5.3** Trends in procurement – percentage of respondents identifying as most often used (NBS 2012, 2015 and 2018).

| Procurement Method | 2012(%) | 2015(%) | 2018(%) |
|---|---|---|---|
| Traditional | | | |
| Client | 59 | 29 | 37 |
| Contractor | 49 | 49 | 43 |
| Consultant | 72 | 37 | 48 |
| Design and Build | | | |
| Client | 26 | 53 | 46 |
| Contractor | 38 | 34 | 33 |
| Consultant | 22 | 52 | 41 |

The NBS does acknowledge other procurement routes such as management contracting and partnering/alliancing, but these are classed as niche within the reports on the basis that less than 3% of the sample identified these as 'most often used'. However, whilst the RICS surveys have considered lump sum, target cost, remeasurement and prime cost (cost reimbursement) procurement routes, the NBS survey labels these *pricing mechanisms*, suggesting they are a tool within the procurement process but not stand-alone procurement routes. Nevertheless, this data can be extracted from the NBS surveys undertaken in 2013, 2015 and 2018, as shown in Table 5.4.

As with the RICS surveys that proceeded the NBS, the data collected reinforces the longitudinal findings that a fixed price or lump sum price mechanism continues to dominate procurement. However, the survey reveals varying take up of other pricing techniques including remeasurement and target cost, although neither method seems to be taking hold within the market. For instance, only 7% of respondents in 2018 identified remeasurement as the pricing method they use 'most often', down from 13% in 2015.

Whilst this data is helpful, it is limited by the sample size and the questions asked. The literature more generally is critical of these approaches to categorisation, as there is an evident desire to place procurement routes into boxes and to give that box a label, be that as a procurement system, pricing mechanism or procurement route. Furthermore, the description of some procurement routes as price mechanisms in the NBS survey adds further confusion. For instance, the RICS Contracts In Use survey considers target cost to be a procurement route whereas the NBS *National Construction Contracts and Law Report* identifies the same technique as a pricing mechanism.

This seeming confusion about what constitutes a procurement route led the British Standards Institution to clarify in BS8534 that all these procurement terminologies are in essence derived from contract pricing mechanisms. As such, they are a mixture of procurement options that can be combined to provide a vast array of methods for delivering the project. For instance, a contractor could be selected from a framework, the project then developed using a design and build procurement route, but supported by an NEC form of contract to facilitate a partnering ethos and high levels of trust and collaboration. Such an approach would seemingly sit across a number of the categorisations outlined in both the RICS and NBS data. With this in mind, in BS8534 the British Standards Institution

**Table 5.4** Trends in pricing mechanism – by frequency of use (NBS 2013, 2015 and 2018).

| Pricing Mechanism | 2013(%) | 2015(%) | 2018(%) |
| --- | --- | --- | --- |
| Lump Sum or Fixed Cost | 64 | 63 | 81 |
| Target Cost | 13 | 17 | 5 |
| Remeasurement | 14 | 13 | 7 |
| Cost Reimbursement | 1 | 2 | 2 |
| Cost Plus Reimbursement | 3 | 1 | 1 |
| Guaranteed Maximum Price | 5 | 2 | 3 |
| Other | 1 | 1 | 0 |

(BSI 2011) advises that procurement decisions should not be made as they traditionally have been by focusing on contract types (which is what the classifications above advocate) and working backwards from that point into strategy. Rather, the professional team should make decisions based on six criteria, which are outlined in Table 5.5, with procurement methods selected based on how the project fits within the scope of each variable and not based on preferences from the past such as 'we have always used D&B so why change' or subjective perceptions about the best procurement route.

### 5.7.2 Pricing Mechanism

As a precursor to the discussion of procurement routes, it seems eminently sensible to first look at the pricing mechanisms that can be used to support the various procurement routes. The latest NBS survey data (NBS 2018) presented in Table 5.4 reveals that lump sum or fixed-cost pricing dominates the construction landscape with other approaches to project pricing seemingly rarely used. Whilst this can be seen as a risk-based decision, with fixed-cost pricing offering a greater level of financial certainty. For some maintenance activities, the approach to pricing needs the flexibility that comes with some of the other techniques identified in Table 5.4.

**Table 5.5**   Procurement decision criteria (Adapted from Hughes et al. 2015, p. 105 and BSI 2011, p. 13).

| Criteria | Considerations |
| --- | --- |
| Source of Funding | The source of funding might not be an option as it can be a matter of policy or regulation for any given client. Sources can include equity financed (owner, developer, government), debt financed (project or organisational level with loan agreements or sale of bonds, etc.) and finally other models such as PPP (Public–Private Partnerships) where the private sector finance the project. |
| Selection Method | The methods of selection should be based on an assessment of relative value and risk to the client. Methods of selection can include negotiation, competitive tender via open or selective methods, partnering, framework agreements. |
| Price Basis | Clients should ensure the price basis reflects earlier decisions about risk apportionment. The prevailing market conditions should be taken into account when deciding on the price basis. Abnormally low bids should not be automatically accepted. Pricing mechanisms will be discussed in more detail later but can include fixed price, target cost, cost reimbursement, etc. |
| Responsibility for Design | The question about the involvement of designers independent of the contractors/suppliers should be resolved based on the extent to which the client's design exigencies take priority over other constraints on the procurement. Design options include client retaining some or all of the design, a novated approach within a design and build or most design works completed by the supply chain. |
| Responsibility for Management | Clients should decide whether to retain responsibility for management of the contract through the construction process either by using in-house teams or consultants. Responsibility for management and co-ordination of site activities can be transferred to a main contractor (under most procurement approaches) or to a construction manager (contractor). |
| Supply Chain Integration | The focus here is on the culture of the project and the extent to which the client feels a trust and collaboration ethos should embed in the supply chain. Options available are integrated, collaborative, fragmented (typical in the sector), competitive. |

**Fixed Price and Lump Sum**

As noted above, this has been identified as one of the most popular approaches to contract pricing, with both the RICS and NBS surveys revealing its dominance of the marketplace. A fixed price arrangement is one whereby the price is agreed and fixed before the contract is signed, often with tenderers asked to bid based on their fixed price for the project. The agreed price is then paid regardless of the contractor's actual costs for the delivery of the project. Generally, the use of fixed price contracting has been seen as a foundation for the adversarial culture as the risk of price fluctuations are taken by the main contractor and passed into the supply chain. Although there is scope within the contract to allow for varying levels of price movement risks to be retained by the client, these are seldom used.

Whilst the notion of fixed price would suggest 100% price certainty, this is seldom the case, as with all projects client change and unexpected problems associated with refurbishment and capital maintenance projects will often give rise to change via variations (JCT) or compensation events (NEC); this will inevitably lead to price increases as the project develops. However, as discussed in Chapter 9, a robust risk management process will ensure the financial impact of such events had been predicted and suitable allowances made by the client in their financial costings for the project.

**Cost Reimbursement**

This payment mechanism allows the contractor to recover all their expenditure on the project from the client, within the scope of a series of strict safeguards. Often referred to as actual prime costs or in NEC allowed costs, the contractor can claim the costs of labour, materials and plant used from the project on the basis of open book accounting whereby the contractor submits evidence in support of their monthly claim. In addition to the prime costs, the contractor will also be paid a pre-agreed percentage or fixed fee to cover their overheads and profit. The clear disadvantage of this approach to the client is the lack of financial control and the seeming lack of incentives for contractor to enact any form of financial management or financial control to keep expenditure as controlled as possible. For these reasons, this approach to pricing will only be used in urgent situations or where the project is relatively small and therefore costs cannot escalate out of control too quickly. A modified and less risky approach to cost reimbursement is to adopt the target cost pricing mechanism, discussed in the next section.

**Target Cost**

The concept of a target cost is actually a misnomer, because there is no such thing in the contracts as a target cost. However, despite this, the notion of a target cost is a well-used term often associated with the NEC option C and D forms of contract amongst others. Favoured by the public sector, given target costs alignment with the constructing excellence and policy agenda, the use of target costs in the planned maintenance sector, especially in social housing organisations has become common place. The notion of a fixed 'target' price provides clients with some level of cost certainty whilst the prospect of financial bonuses for beating the target empower main contractors with lever in value, through their review of the production process and modifications to the design and specification to enhance buildability and improve efficiency. Furthermore, the use of a target mechanism enhances collaborative working and develops a trust culture between the client and the main contractor that is highly regarded in public procurement policy since its initial proposal in the 1994 Latham report. Critics of the method, however, argue the use of a cost target results in the project values being inflated as contractors seek to protect themselves from post-contract risk. Consequently, the target is set so high that it covers most eventualities and by virtue leads to cost savings and thus bonus payments to the main contractor. A simple example is provided below illustrating the operation of this form of contract.

### Scenario

The refurbishment of an office building has been let using a target pricing mechanism. The agreed 'target' was set at £20m; however, the final cost of the works is £18.5m giving a total saving against target of £1.5m. This now needs to be shared between the parties.

The contract stipulates:

| Share Range | Contractor's Share Percentage | Client's Share Percentage |
| --- | --- | --- |
| Less than 95% | 40% | 60% |
| From 95% to 105% | 50% | 50% |
| From 105% to 110% | 70% | 30% |
| Greater than 110% | 90% | 10% |

The contractor's share of the £1.5m saving can now be established:

| Share Range | Saving in Cost Bracket | Contractor's Share (%) | Contractor's Share |
| --- | --- | --- | --- |
| From 95% to 100% – Savings between 19.0m to 20.0m | £1m | 50% | £500,000 |
| Less than 95% Proportion of contract value below £19.0m | £500,000 | 40% | £200,000 |
| Totals | £1.5m | | £700,000 |

In the calculations above you can see the £1.5m of savings generated have distributed between the target brackets, these work in the same way as the taxation system so the first 5% of savings are placed in the first bracket and the remaining savings are placed in the second bracket. Taking the first savings, 90–100% of the contract value is simply calculated by multiplying the contract value by the percentage range. So £20m x 100% = £20m and £20m x 95% = £19m. In this first bracket, any savings above £19m are included. From the example above, you can see this is £1m. (finished project cost £18.5m). The remaining savings, those between the final price and 95% of the contract value (in our example £19m) are located in the second bracket, so for the example above this is the remaining £500k.

Once the savings are distributed in the table, they are multiplied by the weighting share and that is allocated to the contractors. Therefore, for savings between 95 and 100% the contractor receives 50%. In the table above you can see the £1m saving multiplied by 50% gives the contractor £500,000.

**Remeasurement**

This approach to pricing works is often used in two situations. The first is in relation to civil engineering works, which is outside the scope of this book. The second is in the execution of maintenance work. In this regard, the maintenance work can be capital refurbishment schemes, where a series of structures are to be refurbished over a period of time. In this context, the works will often be let based on building types. For example, if you are letting a contract to install new heating systems in the social housing sector this could be simply house type A, B, C and so on. Each building type or work type will then be priced by the contractor on a fixed basis with the client using this rate to then apply to completed heating systems. Alternatively, remeasurement can be used in the sense of a bill of quantities whereby the works are approximately measured for the propose of letting the contract, then as work proceeds the completed work is measured and payments are made based on the measured quantities using the pre-agreed fixed price rates stated in the contract documents.

**Term Contracting**

Despite not being identified as a price mechanism in either the NBS or RICS procurement and contract surveys, in the maintenance sector, term contracting is a well-used pricing mechanism for the outsourcing of reactive maintenance. Term contracts can be let in two ways:

- Measured term contract – based on an agreed schedule of rates.
- Daywork term contract – let on a cost reimbursement basis.

Although the majority will be let on the basis of a measured term contract with the contract documents taking the form of a priced 'schedule of rates' that capture the full range of potential maintenance actions, for most clients a pre-prepared schedule of rates is adopted. One of the leading providers of such documents is NSR management, which produces a series of national schedules of rates including:

- Building works
- Painting and decorating
- Mechanical services
- Electrical services
- Housing maintenance
- Access and adaptations
- Highways maintenance

Once the schedule of rates is priced and a basis for addressing price fluctuations over the course of the contract is agreed, the prices are then used as a basis for the award of a term contract, which is let based on a given period of time and restricted to a geographical location, such as Northwest England. The contractor will be paid for the various maintenance interventions instructed by the client. This is usually instructed via a maintenance help desk as outlined in earlier chapters.

### 5.7.3 Procurement Routes

Following on from the discussion of pricing mechanisms, the next aspect of the procurement strategy to address is the procurement route to be adopted. The earlier discussion of procurement classification identified a series of potential procurement routes, some of which have been deemed pricing mechanisms and discussed in Section 5.7.2, whereas the remainder, which are procurement route options, will be discussed in this section. Whilst the limitations and problems of viewing procurement through the lens of silos or boxes has been debated in earlier sections of this chapter, for the discussion of procurement routes this mentality becomes essential. Yet the reader is advised to see the boundaries of these boxes as permeable and remember that when making procurement route decisions sometimes a blend of features and attributes will be needed to best fit the overall strategy and project requirements. The NBS suggests two procurement routes dominate UK practice: traditional, and design and build. Other forms of procurement such as management and partnering are then seen as niche. In the discussion below, the various procurement routes identified within the literature are discussed, starting with the dominant approaches before some of the niche approaches are explored.

### Traditional Procurement or Design–Bid–Build

Identified in the NBS survey as one of the most frequently encountered approaches to procurement, traditional procurement represents a form of procurement, depicted in some reports as a 'game of two halves' in that the design work is undertaken, the project is tendered and once the successful contractor is identified, construction works commence. This is the traditional architect-led approach to construction procurement. The client appoints the consultant team, headed by the architect, who takes control of the design and cost management of the pre-contract phase of the project. This normally cover stages 0–5 of the RIBA (Royal Institute of British Architects) plan of work, although it is possible that a limited amount of highly specialist design will be handed to the contractor to complete via a provisional sum and contractor's design portion within the contract. The typical organisational structure for traditional procurement is illustrated in Figure 5.5.

Upon completion of the design works, the client will then tender the project, normally competitively and on a fixed price basis, to identify and appoint a contractor who will

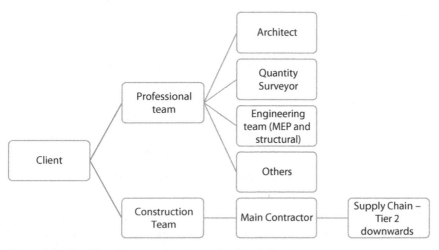

**Figure 5.5** Traditional procurement organisational chart.

undertake the construction or manufacturing phase of the project, including the management of subcontract organisations. The tendering process is typically based on a detailed set of drawings, a supporting specification and some form of pricing document, either a bill of quantities or activity schedule/schedule of work.

Textbooks say that the key to this route is the full completion of design works prior to the project being issued for tender; however, the RICS (2013 p. 8) suggests there are a number of route options existing within the guises of traditional procurement. These include:

1. *Sequential* – As outlined above, this is the traditional perspective, with design works completed prior to tendering. The successful contractor then would proceed with the construction and manufacturing phase of the project.
2. *Accelerated* – Some of the design works overlap with the construction; this is achieved by letting a separate advanced works or enabling works contract, for example, demolition and site remediation works, be undertaken. Or the full groundworks package could be let whilst technical details for the main building are finalised (RICS 2013).
3. *Two-stage tendering* – The final option outlined in the RICS guidance note is the possibility of letting the works in two stages. The first stage allows the contractor to be selected based on predicted costs for major work elements, with the detailed pricing negotiated at the second stage once design has concluded. This allows the client the option of enhanced buildability by having some contractor involvement in the design of the project.

Traditional procurement offers a number of benefits to the client, including:

- Competitive and fair bidding, as all contractors have the same information at the tender stage of the project.
- The client retains a lot of design influence and can inform the direction of the project in terms of performance, aesthetic and other quality requirements.
- The fact that the design is almost complete at the point of tender affords the client a high level of financial control and cost certainty. Post-contract change is inevitable with all forms of procurement, however, especially in the maintenance and refurbishment sector where the full condition of the existing building may not be fully ascertained until works commence.
- Where public expenditure is concerned, traditional procurement ensures accountability and transparency with appropriate levels of competition.
- The approach to procurement is well understood and procedures for the management of the project are clearly outlined, giving confidence to the supply chain.
- Changes are reasonably easy to make post contract with high levels of financial and design information ensuring reasonably easy change management processes.

Despite the many benefits, traditional procurement also presents a number of drawbacks to the client. For example:

- If the client tries to accelerate the process by issuing incomplete design information this can result in extensive post-contract change, removing time and cost certainty from the project.

- Traditional procurement takes a long time due to the sequential nature of the two main stages (design and construction).
- Late contractor involvement is at the core of this approach to procurement. The issues with this are well highlighted and widely criticised in the literature due to the lack of opportunities for embedding buildability and the enhanced value associated with early contractor involvement.
- The focus on the approach with design and construction completely separated is seen to encourage adversarial relationships.
- For certain types of projects, especially those in the maintenance and refurbishment sectors, project scope can be difficult to fully define, so the use of traditional procurement is likely to result in the client paying an enhanced risk premium.
- Both designer and contractor have little incentive to advise the client on factors that may benefit the operational costs of the completed building, thus diminishing the value benefits that could be achieved during the design and construction of the project.

### Design and Build

The second dominant procurement route identified in the NBS survey (NBS 2018) is the design and build approach. Unlike traditional approaches to procurement, design and build is a relative newcomer. Design and build as we understand it today was invented by contractors in the early 1960s when a number of contracting organisations, likely seeking to gain some competitive advantage in the marketplace, began to offer clients a package deal. Contractors would offer the client the complete package of works including design and construction in place of the comparatively fragmented traditional arrangement (Boudjabeur 1997). The design and build organisational structure is outlined in Figure 5.6.

Buoyed up by the increasing complexity of buildings, the need for accelerated project delivery and the scathing findings of the Emmerson Report in 1962, design and build (D&B) began to be adopted in greater and greater proportions by clients attracted by the risk transfer and accelerated delivery opportunities along with the benefits of having a single-party responsibility for both the design and construction. Design and build grew extensively in the intervening years, with Boudjabeur (1997) suggesting that 15% year-on-year increases in the use of the procurement method were being witnessed between the late 1960s and the 1990s. Despite forecasts suggesting that at least 50% of projects

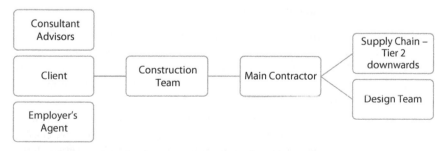

**Figure 5.6** Design and build procurement organisational chart.

would be procured using D&B by the turn of the twenty-first century (Boudjabeur 1997), trends in the design and build adoption have levelled off, as the latest use data from NBS suggests 41% of respondents identified design and build as their most used procurement approach (NBS 2018). Even so, design and build remains an important procurement route for project delivery in the UK, given the high levels of risk transfer, single point responsibility for latent defects and accelerated project delivery this approach to procurement offers.

The RICS guidance note *Developing a Construction Procurement Strategy and Selecting an Appropriate Route* (RICS 2013) identifies a number of variants of design and build procurement that the client can utilise, the most common of these is an approach which affords the client some level of project control, often with tendering occurring at RIBA stage 3. In this approach to design and build the client will employ a design team to carry out some preliminary design work and to prepare a project briefing including detailed *employer's requirements*, (a term used in some contracts to describe the basis for the contractor's submission). The client's requirements are likely to include project objectives, significant criteria for a successful outcome. This information could include items such as performance specifications, layout requirements or potentially room data sheets giving very specific requirements for each space within the building. Other forms of design and build include:

- *Turnkey or package* – Minimal client involvement, whereby the contractor is left to interpret the client's requirements and provide a building as a complete package. This can in some instances include seeking planning permission or potentially assisting with the project funding.
- *Develop and construct* – A strategy whereby the client appoints designers to prepare the concept design before the contractor assumes responsibility for completing detailed design and the construction of the asset. In this form, the contractor develops the design from a detailed brief or performance specification and periodically updates the client on design development, allowing the client to ensure the contractor's interpretation of the brief meets their requirements.
- *Novated* – In this approach, the successful contractor assumes responsibility for the design team appointed by the client and is responsible for the initial design development usually to stage 3 of the RIBA plan of work. This allows the design team to continue to develop the project design. However, contractually, their client changes from the main client to the main contractor. Regrettably, this is sometimes resisted by the design team who raise ethical and other challenges.

Whilst design and build offers the client a number of benefits, as outlined earlier these tend to relate to risk transfer possibilities into the supply chain, accelerated delivery when compared with traditional forms of procurement and, importantly, a single point of legal responsibility for both design and construction. Yet it is this third point that has proven to be highly contentious. Under conventional professional service contracts associated with architect-led design, the architect affords the client a level of legal obligation termed 'reasonable skill and care' as required under the Supply of Goods and Services Act 1982. The common law test of negligence, in this context, is one which 'provides that a professional person is not negligent if he or she carries out their work to the same standard that another reasonably competent member of their profession would have met' (Fenwick Elliot 2015).

However, within design and build, the contractor is failing a dual role: at one point they are providing a service, in that they are developing the design of the building, whilst later on they are constructing the building so become producers. As such, Fenwick Elliot (2015) argue the constructor 'could be said that he is under conflicting obligations in respect of the two distinct functions of design and construction'. As a result, most clients have sought to hold the design and build contractor to the higher duty of *fitness for purpose*; this is an 'absolute obligation to achieve a specified result, a breach of which does not require proof of negligence. This duty stems from the Sale of Goods Act 1979 which imposes implied terms on any seller acting in the course of business that the goods supplied will be of satisfactory quality and, where the purchaser makes known any particular purpose, are reasonably fit for their intended purpose' (Fenwick Elliot 2015). Although contract drafters such as the JCT have taken steps to remove this absolute duty and move to the lesser obligation of reasonable skill and care, there remains a pressure from clients to seek to reimpose the higher-level duty through contract amendments (RICS 2013).

Despite these legal challenges to design and build, the route offers a number of benefits. For example:

- The client achieves a single point of responsibility with only one firm to deal with. This can also significantly reduce client expenditure in terms of professional consultants.
- Aggressive risk transfer is facilitated through this approach to procurement due to the single point of responsibility for both design and construction.
- Early contractor involvement in the project can enhance value through improved buildability and production processes.
- Price certainty is generally obtained prior to the commencement of construction.
- The overall project time frame for design and construction is reduced, as these stages are intertwined and overlapping rather than linear and sequential.

Once again, despite the many benefits, design and build also presents a number of drawbacks to the client, and these include:

- The production of detailed employer's requirements to allow control of performance and quality can be difficult, especially if the full requirements of the asset are not fully understood.
- There needs to be a willingness from the perspective of the client to commit to design concepts at the early stages of the RIBA plan of work.
- Tenders from different contractors can be difficult to compare as design development is likely to have taken place.
- Client change is not really encouraged and if client changes are made, they can be expensive and often difficult to price due to a lack of detailed cost information.
- Design liability can be limited depending on the contract used and conditions amended.
- Quality is often relinquished to the contractor, making client control difficult.
- Whilst buildability and space performance is enhanced, this often comes at the price of the design aesthetic.

**Management Procurement**

Despite the surveys of procurement approach defining the management routes as niche, they still pay an important role in the construction process and as such they are discussed here. Whilst it would not be expected that these procurement approaches feature in the asset management field often, with only large-scale and complex projects adopting the route, there are exceptions. For instance, the £236m refurbishment of the grade 1 listed Manchester Town Hall is currently on-site, having appointed Lendlease using one of the management procurement routes (Manchester City Council 2019).

There are two approaches associated with management procurement: construction management and management contracting.

**Construction Management**

Construction management is the procurement route chosen recently on several high-profile projects, including Tottenham's new stadium and in the asset management field, the Manchester Town Hall refurbishment project. Under construction management, the client does not allocate the risk and responsibility for project delivery to a single organisation. Instead, the client employs the design team but also appoints a contractor to act as 'construction manager'. This appointment tends to be made on a fee basis with the construction management contractor taking responsibility for planning and scheduling the works, including the co-ordination of a myriad of work package contractors with the construction manager taking responsibility for the timely completion of the overall project. The organisational chart shown in Figure 5.7 illustrates the complexity of the contract arrangements. Whilst the construction management contractor is responsible for project delivery, the contracts for the work package contractors are let by the client and with no contractual links created back to the construction management contractor, only a functional control relationship is created. As a result, the client will often need to employ a project management specialist to oversee the work of the construction manager and to deal with various administrative processes, such as making interim payments to the various work package contractors.

As mentioned earlier, the construction management as a procurement route tends to be adopted for large-scale and complex projects where the client's primary objective is to accelerate project delivery without as much focus on financial control. This is important

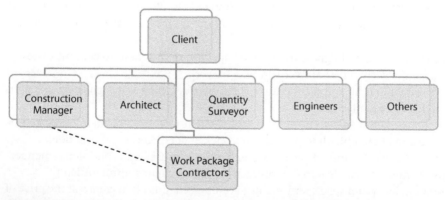

**Figure 5.7** Construction management procurement organisational chart.

as construction management offers little in terms of cost certainty until all the work packages have been fully designed and let.

### Management Contracting

The second approach to management procurement is management contracting, the RICS (2013) suggest this was once a popular route for management procurement in the 1980s and 1990s but has since declined in popularity in favour of construction management. The premise is the same, in that a contractor is appointed to oversee the works from a management perspective. However, unlike construction management, under management contracting the contract is engaged to manage the whole of the construction process. The contractor will be paid a fee on top of the prime costs of construction (based on work packages) to undertake this oversight role, with the fee determined based on the quantity surveyor's cost plan estimate cost for the project. The work packages are let to 'works contractors' and, unlike with construction management, the contractors have a direct contractual link to the management contractor who takes full responsibility for the administration of the various works contracts. This is illustrated in the organisational structure provided in Figure 5.8.

Herein lies the problem: issues with late payment or non-payment by the management contractor would often lead to project delays and would often use client payments to engineer positive cash flows within the business to reinforce other projects. Other accusations levelled at management contractors included unreasonable and commercially adversarial reductions in valuations using the rules of set-off (now known as pay less notices) to dispute works completed, or to falsely claim the works did not meet the client's quality requirements and rework would be needed (Higham et al. 2016, p. 88).

### Partnering and Frameworks

The term 'partnering' become fashionable in the UK construction sector in the 1990s and remains so to this date. Initially proposed through the publication of two seminal reports, *Constructing the Team* by Sir Michael Latham in 1994 and the later publication *Rethinking Construction* by Sir John Egan in 1998, partnering or collaboration is still the cornerstone of

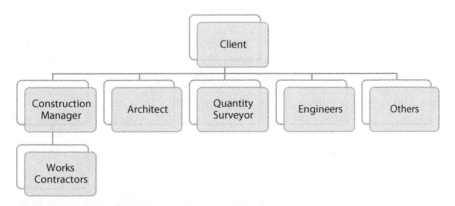

**Figure 5.8** Management contracting organisational chart.

construction policy with the construction sector deal (HM Government 2018) due to the benefits created in terms of greater efficiency and benefits of creating groups of trusted parties.

Both reports identified partnering and collaboration as a future direction for the sector to enhance project satisfaction and reduce the adversarial nature of the marketplace. Partnering, as defined by the RICS (2013, p. 17), is 'a cooperative relationship between business partners formed in order to improve performance in the delivery of projects'. Yet most literature is of the view that partnering, although often defined as a procurement route, is not in itself a unique procurement route. Rather, partnering is 'best considered as a set of collaborative processes which emphasise the importance of common goals and raise such questions as how such goals are agreed upon, at what level are they specified and how are they articulated' (RICS 2013, p. 17). In this way, partnering can be applied to either a single project, known as project partnering, or it can be applied in the form of a longer-term relationship over multiple projects, such as in the case of a framework agreement. In this longer-term form, partnering becomes known as either strategic partnering, or if in the rail sector, alliancing (Designing Buildings Wiki 2020).

Current levels of collaboration within the construction sector can be appraised via the NBS *National Construction Contracts and Law Report* published in 2018 (NBS 2018) which sought to survey respondents on their current levels of collaboration. The findings revealed a mixed picture, with only 16% of respondents suggesting some level of collaboration was achieved on all projects they are involved with. However, more promisingly, 63% suggested they used collaboration on some projects, with only 37% saying they never engaged with any form of partnering or collaboration. Despite this, the survey revealed high value projects were more likely to see collaboration (78%). Even so, over half (57%) said they had experienced collaboration on low value projects, such as those more akin to the refurbishment and maintenance sector. Importantly, the authors of the report concluded this finding suggested collaboration and partnering was 'as much about the ethos of the organisation as the value of the project'. Importantly for the refurbishment and maintenance sector, especially when it comes to outsourcing maintenance, the organisations normally embed themselves deeply in a collaborative relationship with joint logos on vehicles and employees' workwear, etc.

Although when asked how collaboration was facilitated, the majority of respondents identified contract conditions that included an ethos of 'mutual trust and cooperation' to be the most common tool for facilitating collaborative working. With only a third of the respondents suggesting a more structured approach via the adoption of a formal partnering agreement had been used.

Despite partnering not being seen as the stand-alone procurement route it is often claimed to be in some textbooks, partnering does offer a number of benefits when interwoven with other procurement routes outlined in this section. These include:

- A reduction in post-contract dispute and a lessening of adversarial practices.
- Early supply chain involvement if integrated supply chains are used with a collaborative ethos.
- It is based on a win-win culture and adopting an open book approach to financial issues.
- Design and Construction integration are possible, especially with very early contractor involvement.
- It allows benefits associated with a trust-based collaborative working relationship to be extended across multiple projects.

Yet, as the relatively low exclusive uptake evidenced above demonstrates, partnering also brings some disadvantages, including:

- The partnering and trust-based ethos associated with the process can be abused by one of the parties.
- The process requires more client resources to compensate for the less competitive environment and the process can collapse if one party becomes too disadvantaged.
- The partnering process is a major culture shift for many in the sector and so can take years to perfect.
- An early commitment is required in terms of management resources and direct costs.
- There are the direct costs of workshops, of training staff, and of the more intensive early involvement of management in establishing the partnering approach.

## 5.8  Contract Arrangements

The final area for exploration in this chapter is the contractual arrangement or terms of engagement used for used for the delivery of refurbishment and maintenance programmes through the guise of appointing the project team. This will invariably include the professional consultants, main contractor and their supply chain. that are. As this book is not exclusively focused on the administration and management of these contracts, this section will focus on some of the more specialist forms of contract aimed at the reactive maintenance and potentially planned maintenance market that are seldom mentioned in other texts looking at construction contracts.

More generally, the National Building Specification (NBS 2018) conducted the *National Construction Contracts and Law Survey* of employers, consultants and contractors over the period August to November 2017. Based upon 360 responses, it reported that the most popular contract between employers and contractors is the Joint Contract Tribunal (JCT) suite, accounting for 62% of contracts used by their respective organisations, whilst the New Engineering Contract (NEC) accounted for 14% of contracts. Other forms of contract including Bespoke, FIDIC, PPC2000 and others made up the remaining 24%. A full breakdown of this data is provided in Table 5.6. The survey discovered 70% of respondents had used the JCT contract in the last 12 months (increased

**Table 5.6**  Contracts used in last 12 months (Adapted from NBS Survey 2018).

| Form of Contract | 2018(%) |
| --- | --- |
| JCT Contracts | 70 |
| NEC Contracts | 39 |
| Bespoke Contracts | 23 |
| RIBA Contracts | 14 |
| FIDIC Contracts | 10 |

from 57% in 2015 survey), whereas only 39% had used the NEC within the same time period. Of more interest, especially when looking at the refurbishment and maintenance market, is the fact that the survey revealed that contract choice aligns closely with project value. The survey reveals the universal popularity of the JCT form of contract, with 25% of respondents adopting JCT for projects between £50,000 and £250,000 and 47% for projects between £250,000 and £5m, whereas the NEC was typically favoured for projects with a contract value (price) between £250,000 and £5m, and £5m and £250m (NBS 2018). However, the survey revealed little as to the alternatives used at the lower end of the market which is likely to be dominated by the refurbishment and maintenance sector.

Given the dominance of the JCT and NEC in the UK construction market, this chapter will conclude by providing a general overview of the two contract suites. It is not intended in this text to provide a detailed commentary on any of the specific contract types, as this is well covered elsewhere in contract textbooks and would require extensive additional chapters.

### 5.8.1   The Joint Contracts Tribunal (JCT) Suite of Contracts

The most popular and most adopted forms of contract in the UK fall within the umbrella of the JCT, the Joint Contracts Tribunal. Established around 1870, the JCT is the outcome of a series of professional and other trade bodies in partnership with the Royal Institute of British Architects in the UK discussing the possibility of creating a standard form of contract for use in the construction sector. Initially intended as a response to concerns about the applicability of general common law contractual provisions and the expense of bespoke contract drafting, the use of a standard form, it was hypothesised, would offer a quick, cheaper and effective way of establishing a common contract for the sector. The first attempt at drafting a standard contract was made in 1903, but it was not until 1909 that the contract we now know as JCT emerged as a common agreed document.

At the current time, the Joint Contracts Tribunal's membership includes a number of professional and trade bodies representing the various facets of the construction sector including:

- Royal Institute of British Architects
- Construction Confederation
- Royal Institution of Chartered Surveyors
- Local Government Associations
- Association of Consulting Engineers
- British Property Federation (Joining in 1992)
- Scottish Building Contract Committee
- National Specialist Contractors Council

Each organisation becomes a member of one of five colleges within JCT that represent the interests of:

- Employers, clients and local authorities
- Consultants
- Contractors
- Specialists and subcontractors
- Scottish building industry

This provides the JCT with a unique structure amongst contract drafting bodies, as it can be said to represent the perspectives of both the demand (clients and their consultants) and supply (contractors and subcontractors) sides of the construction market; whereby consensus between the two sides on the structure and content of the JCT form of contract, in its unamended form, is said to ensure an even and equitable distribution of risk. However, in practice, it is well known that most contracts will undergo amendment prior to tender to rebalance the risk in the favour of the employer and to transfer as much risk as possible to the main (tier 1) contractor and beyond into the supply chain.

Despite the popularity of JCT, the suite of contracts has undergone significant revisions over the last 20 years. In 2005, the suite was substantially revised with the intention of improving accessibility and to ensure a uniform layout across the suite with the end outcome being a far more user-friendly suite of contracts (Thompson and Price 2010). A second, equally comprehensive series of revisions took place in 2011 in response to changes to payment provisions. The contract was updated with the new provisions introduced into the Housing Grants, Construction and Regeneration Act 1996, and the accompanying Scheme for Construction Contracts, arising from Part 8 of the Local Democracy, Economic Development and Construction Act 2009 which came into force in late 2011 (Knox 2011). At the same time, a series of other changes were introduced to capture issues such as site waste management plans, dispute resolution, insurance provisions, some refinements to relevant events (extension of time) and termination/insolvency provisions; whilst a relatively minor series of changes introduced in 2016 focused on further enhancing the suite's functionality and user-friendliness. Specific amendments included revisions to the provisions around payments to align with the Fair Payment Charter, and a general update of the suite for changes in statute provisions and the incorporation of the JCT Public Sector Supplement into the main contracts (Atkinson and Wales 2017).

The major JCT forms are shown in Table 5.7, which summarises the appropriate forms to be used in various circumstances.

### 5.8.2 The New Engineering Contract (NEC) Suite of Contracts

The NBS (2018) contract and law survey results identified the NEC suite of contracts as the second most adopted forms of contract in the UK, largely due to the dominance of the NEC in the public sector, following its praise in the Latham report and the decision by the UK government to endorse its use for public sector procurement in the Construction Strategy (Cabinet Office 2011). Subsequent decisions to abolish both the ICE (Institute of Civil Engineers) and GC Works (UK Government contract), has led to the NEC enjoying a period of substantial growth and expansion in the UK for use in both civil engineering and public sector projects. The NEC was initially introduced in 1993, as a retort to the seeming lack of project management in JCT amongst other contracts. Martin Barnes, the lead drafter of the NEC commented that 'Our philosophy was to produce something which cured every known ill of traditional contracts. We did not have to compromise. Everything we thought would be a good idea went in – and we could decide what to put in solely on the basis of what would stimulate all those using it to manage their contribution well' (NEC 2021). The NEC focused on collaborative working, the use of plain English and the desire to stimulate, not frustrate, good management. Although critics of the contract suggest this desire to stimulate good project management was taken too far and

**Table 5.7** Selecting the appropriate JCT form of contract (Adapted from College of Estate Management 2014 and JCT 2017).

| Designer of the Works | Contract Documents | Value Range | Recommended Form of Contract | Comments |
|---|---|---|---|---|
| Architect or another consultant, for example an asset manager, building surveyor, architectural technologist, etc. | Drawings Specifications Bill of Quantities | Normally over £1m or reasonably complex | SBC/Q 2016 Standard Building Contract with Quantities | Where some parts of the work are to be designed by the contractor or specialist subcontractors employed by the contractor |
| | Drawings Specifications Approximate Bill of Quantities | | SBC/AQ 2016 Standard Building Contract with Quantities | |
| | Drawings Specifications Activity Schedule or other pricing document | | SBC/XQ 2016 Standard Building Contract with Quantities | |
| Architect or another consultant | Drawings Specifications Pricing Document – Bill of Quantities | £100,000 to £10m | IC 2016 – intermediate form of Contract | |
| Part architect/ part contractor | Approximate Bill of Quantities Activity Schedule or other Schedule | | ICD 2016 – intermediate form of Contract with contractor's design | Allows some limited use of contractor's design |
| Architect or another consultant | Drawings Specifications Pricing Document – | Up to 250k | MW Minor Works Contract | |
| Architect or another consultant with a small amount of contractor design | Activity Schedule or other Schedule | | MWD Minor works with contractor's design | |
| Asset manager/ building surveyor or other consultant | Schedule of Rates | Not stated | RM 2016 – Repair and Maintenance Contract (commercial) | No contract Administrator. For work related to repair and maintenance of a building |
| Asset manager/ building surveyor or other consultant | Schedule of Rates | Any value | MTC 2016 – Measured Term Contract | Periodic repair and maintenance – reactive maintenance |

*(Continued)*

**Table 5.7** *(Continued)*

| Designer of the Works | Contract Documents | Value Range | Recommended Form of Contract | Comments |
|---|---|---|---|---|
| Architect or another consultant | Drawings (possibly) Brief specification Cost estimate/ Cost plan | Any value | PCC 2016 – Prime Cost Building Contract | Cost reimbursement projects where an early start is essential |
| Contractor | Employers requirements Contractors proposals Cost analysis/ Cost plan | Any value | DB 2016 – Design and Build Contract | See discussion earlier in the chapter |
| | | £100k to £100m | MP 2016 – Major Project Construction Contract | Client and contractor experienced in project type and both have established contract procedures in place |
| Architect or another consultant | Drawings Specification Prime Cost of Project/Cost plan | Over £1m | MC2016 Management Building Contract CM/A Construction Management appointment | Complex projects requiring management expertise. Where early start is required or acceleration of the works |
| | | Any Value | CM/TC Management Trade Contract | Used with Construction Management for appointment of works package contractors |
| Architect or another consultant | Drawings Specifications Pricing Document – Bill of Quantities, Approximate Bill of Quantities, Activity Schedule or other Schedule | Any Value | CE 2016 – Constructing Excellence Contract | For projects where participants wish to engender collaborative working |
| | | | CE/P Constructing Excellent Contract Project Team Agreement | Engenders collaborative approach and formalises the integration of the project team / Can include risk and reward sharing between team members |
| Architect or another consultant | None | Any Value | FA 2016 – Framework Agreement | For clients who carry out work regularly and wish to try and capture the benefits of long-term relationships within the supply chain |

*(Continued)*

**Table 5.7** *(Continued)*

| Designer of the Works | Contract Documents | Value Range | Recommended Form of Contract | Comments |
|---|---|---|---|---|
| Part architect or another consultant/part contractor | Drawings Specifications PricingDocument – Bill of Quantities, Approximate Bill of Quantities, Activity Schedule or other Schedule | Any Value | PCSA 2016 – Pre-Construction Service Agreement | For the supply of pre-construction services by a contractor selected under a two-stage tendering procedure |
| Homeowner | Drawings Specifications | Not stated | HO/B Building contract for homeowner/occupier | No consultant appointed |
| Consultant | | | HO/C Building contract for homeowner/occupier | Consultant appointed to oversee works |
| Homeowner | Specifications | Not stated | HO/RM Home repair and maintenance contract | No consultant appointed |

created an administratively burdensome contract, as exclaimed by Mr Justice Edwards-Stuart, who in the case of *Anglian Water Services Ltd. v Laing O'Rourke Utilities Ltd.* [2010] EWHC 1529 (TCC) described the contract as a 'triumph of form over substance'.

When initially developed between 1991 and 1993, the NEC suite provided a single contract for both building and civil engineering. The contract was revisited in 1995 to incorporate the findings of Sir Michael Latham's *Constructing the Team* report to the UK government, leading to the publication of the second edition, called the NEC Engineering and Construction Contract (ECC), although this time the suite had been expanded to also include a professional services contract, adjudicator's contract and a back-to-back series of short forms and subcontract forms of contract. A further, substantially enhanced form of the contract was published in 2005, the third edition or NEC3, once again provided a wider range of options within the suite including a term service contract, framework contract and supply contract. Finally, in 2017, the fourth edition of the NEC4 was published, updating the contract suite to meet changes in statute provisions and emergent case law. However, the publication of the fourth edition also allowed further expansion of the contract suite with a new PPP (design, build, finance and operate) contract and an alliancing contract introduced.

The NEC4 suite of contracts for use with the main contractor include:

- Engineering and Construction Contract (ECC)
- Engineering and Construction Short Contract (ECSC)
- Engineering and Construction Subcontract (ECS)
- Engineering and Construction Short Subcontract (ECSS)
- Term Service Contract (TSC)

- Term Service Short Contract (TSSC)
- Framework Contract (FC)
- Design, Build and Operate Contract (DBOC)
- Alliance Contract (ALC)

In addition, as with the JCT suite, a series of additional contracts exist for use within the supply chain or for the appointment of consultants. These include:

- Engineering and Construction Subcontract (ECS)
- Engineering and Construction Short Subcontract (ECSS)
- Professional Service Contract (PSC)
- Professional Service Short Contract (PSSC)
- Dispute Resolution Service Contract (DRSC)

Unlike the JCT, the main NEC form of contract (Engineering and Construction Contract (ECC)) operates using a series of nine core clauses, which the client can then add to using primary and secondary option clauses to design the contract to fully meet the requirements of their project. The main primary contract options include:

- Option A – Priced Contract with Activity Schedule
- Option B – Priced Contract with Bill of Quantities
- Option C – Target Contract with Activity Schedule
- Option D – Target Contract with Bill of Quantities
- Option E – Cost Reimbursable Contract
- Option F – Management Contract

These are then supported by a series of 22 secondary X clauses, a series of W and Y clauses designed to address UK or international context of the project, and finally a series of Z clauses, where the client can add bespoke obligations into the contract. These are shown in Table 5.8. Additionally, Table 5.9 maps the various NEC main contract forms to various procurement routes and circumstances under which contract selection may take place.

### 5.8.3   The ACA PPC 2000 Form of Contract

Often used in the social housing sector for asset management and capital works related to refurbishment, PPC2000 is a reasonably niche contract, not appearing in the NBS survey, for instance. Yet it is, nevertheless, worthy of mention here, especially if readers of this text find themselves operating in the social housing sector, amongst others. Written by solicitors Trowers and Hamlins LLP with the input of contractors, lawyers, engineers, architects and others, and published by the Association of Consultant Architects (ACA), the suite of contracts is aimed squarely at those seeking to fully embrace the ethos of collaborative working.

It was introduced in September 2000 as a direct response to Sir John Egan's calls for a shift away from the adversarial contract practices often associated with the more established forms of contract, especially JCT. The PPC suite of contracts presented a major change in construction contract practice, with all the forms adopting a multiparty focus whilst also binding all parties to a partnering agreement that spelt out their responsibilities, not just to the client but to each other. Learning from the project management focus

**Table 5.8** NEC Secondary X, W, and Y clauses (Adapted from NEC 2017).

| Secondary Clause | Main Features |
| --- | --- |
| *X Clauses* | |
| X1 Price adjustment | Transfers risk of price changes to client |
| X2 Changes in law | Transfers risk of changes to common, civil or statute law to client |
| X3 Multiple currencies | Transfers currency exchange risk to client |
| X4 Ultimate holding company guarantee | In others words a parent company guarantee. Affords the client additional protections |
| X5 Sectional completion | Enables client to phase the project and have works starting or ending at different times with contractual control |
| X6 Bonus for early completion | Contractor paid a bonus for each day the project is completed ahead of the contractual completion date |
| X7 Delay damages | Allows client to recover liquidated and ascertained damages from the contractor |
| X8 Undertakings to the client or others | Requires subcontractor or supplier to provide undertakings – collateral warranties to client |
| X9 Transfer of rights | Transfers rights to material prepared for design to the client |
| X10 Information modelling | Address provisions needed to support use of building information modelling (BIM) |
| X11 Termination by client | Widens scope of termination by client to include 'for any reason' |
| X12 Multiparty collaboration | Collective incentivisation of the full supply chain to achieve project objectives. Cannot be used with X20 |
| X13 Performance bond | Additional protection available to client via third party financial surety |
| X14 Advanced payment to contractor | Enables payments to suppliers such as suppliers of off-site manufactured components – does allow for protections via security bond |
| X15 Contractor's design | Limited liability to 'reasonable skill and care' – See D&B discussion for further information |
| X16 Retention | Allows the client to retain a portion of interim payments as security – Usually 3–5% |
| X17 Low performance damages | Contractor pays liquidated damages for defects in design or construction if defect not corrected and becomes listed on defects certificate |
| X18 Limitation of liability | Limited various liabilities afforded to the client |
| X19 Termination by either party | Enables either the client or contractor to terminate the contract for neutral events (frustration in law) |
| X20 Key performance Indicators | Allows client to introduce a range of KPIs for contractor to meet – comes with a financial bonus for attainment |
| X21 Whole life cost | Contractor encouraged to change scope of works if reduce life cycle costs (see Chapter 7 for full discussion of concept) |

*(Continued)*

**Table 5.8** *(Continued)*

| | |
|---|---|
| X22 Early contractor Involvement | Allows contractors to be appointed early to help with design development. Only available with option C and E |
| X23 Extending service period | Can extend period covered by the term service contract |
| X24 Accounting periods | Enables client to periodically achieve financial closure (used with term service contract) |
| X25 Supplier Warranties | Requires contractor to provide warranties to client (used with NEC supply contract) |
| X26 Programme of work | Client flexibility to add further works to scope of contract (used with alliancing contract) |
| ***Y Clauses*** | |
| Y(UK)1 Project Bank Account | Requires contractor to establish and maintain a project bank account for paying named suppliers and subcontractors |
| Y(UK)2 | Aligns contract provisions with UK statute – Housing Grants, Regeneration and Construction Act |
| Y(UK)3 Contracts (rights of third parties) Act 1999 | Allows a third party to enforce a term of contract – can work with or in addition to X8 (Undertakings to the client or others) |
| ***W Clauses*** | |
| W1 | Allows for alternative dispute resolution (ADR) using adjudication, but Housing Grants, Regeneration and Construction Act does not apply (outside UK) |
| W2 | Allows for ADR using adjudication where Housing Grants, Regeneration and Construction Act applies (UK) |
| W3 | Allows for ADR using dispute avoidance board, but Housing Grants, Regeneration and Construction Act does not apply (outside UK) |

**Table 5.9** Selecting the appropriate NEC form of Contract (Adapted from NEC 2017).

| Procurement | Designer of the Works | Contract Documents | Recommended Form of Contract | Comments |
|---|---|---|---|---|
| Traditional – Fixed Price/ Lump Sum Pricing Mechanism | Architect or another consultant, for example an asset manager, building surveyor, architectural technologist, etc. | Drawings Specifications Approximate Bill of Quantities | NEC ECC (Engineering Construction Contract) main Option B | Various levels of contractor design can be added via the core clauses. No scope for firm Bill of Quantities |
| | or part architect/ part contractor | Drawings Specifications Activity Schedule | NEC ECC (Engineering Construction Contract) main Option A | Various levels of contractor design can be added via the core clauses |

*(Continued)*

**Table 5.9** *(Continued)*

| Procurement | Designer of the Works | Contract Documents | Recommended Form of Contract | Comments |
|---|---|---|---|---|
| Traditional – Target Cost | Architect or another consultant for example an asset manager, building surveyor, architectural technologist, etc. or part architect/ part contractor | Drawings Specifications Approximate Bill of Quantities<br><br>Drawings Specifications Activity Schedule | NEC ECC (Engineering Construction Contract) main Option D<br><br>NEC ECC (Engineering Construction Contract) main Option C | Various levels of contractor design can be added via the core clauses. Encourages collaboration via sharing of savings/ overspend |
| Design and Build – Fixed Price/Lump Sum Pricing mechanism | Contractor | Performance Specification Cost Analysis/Cost plan | NEC ECC (Engineering Construction Contract) main Option A | Various levels of contractor design can be added via the core clauses |
| Design and Build – Target Cost | Contractor | Performance Specification Cost Analysis/Cost Plan | NEC ECC (Engineering Construction Contract) main Option C | Various levels of contractor design can be added via the core clauses |
| PPP (Public Private Partnership) | Contractor | Output or Outcome Specification for Asset Performance Life Cycle pricing | NEC Design, Build, Operate Contract (DBOC) | Normally public sector within a PPP |
| Management | Part architect/ part contractor | Drawings Specifications Cost Plan/ Model | NEC ECC (Engineering Construction Contract) main Option F | Applied to all management procurement options |
| Maintenance – Term Contracting | Consultant/asset manager | Schedule of Rates | Term Service Contract (TSC) | Periodic repair and maintenance – reactive maintenance |
| Prime Contracting | Contractor | Output or Outcome specification for asset performance Cost analysis/ cost plan | Can be used with any of the main ECC options | Similar to design and build – design requirements are to deliver the performance requirement for which the asset was intended |

*(Continued)*

**Table 5.9** *(Continued)*

| Procurement | Designer of the Works | Contract Documents | Recommended Form of Contract | Comments |
|---|---|---|---|---|
| Framework Agreement | Various | | NEC Framework Contract (FC) | For clients who carry out work regularly and wish to try and capture the benefits of long-term relationships within the supply chain |
| Alliancing | Consultant team | Various depending on nature of projects | NEC Alliance Contract (ALC) | Multiparty agreement – longer-term collaborative contract with a number of suppliers in order to deliver a large-scale multidiscipline project or programme of work |

seen in the NEC and the form of contract highly praised by Sir Michael Latham in his 1994 report Constructing the Team, the drafters of the PCC contract incorporated similar project management processes. These process are designed to aid the parties with issues such as the partnering objectives, duties of care, key performance indicators, risk, price incentives and change management. However, unlike NEC, the contract did not adopt the same financial incentivisation afforded through the application of a shared pain/gain mechanism, colloquially known as a target contract. For some, the absence of such a mechanism was seen as a major shortcoming of the suite.

As with other collaboratively focused contracts such as the NEC, PPC2000 includes a number of the same essential features that aid and enshrine successful partnering or collaboration, including:

- The formation of a core group of stakeholders, focused on working together to achieve a common series of objectives on the basis of a win-win approach.
- The operation of an early warning system that proactively seeks to identify problems and address those via collective decision-making.
- Early agreement of the party's costs above prime (actual costs) such as profit, site and head office overheads.
- Encouragement within the team to seek value solutions that may reduce overall project costs – although for some this is compromised by the lack of the target mechanism.
- Performance measurement linked to contractual KPIs and explicitly encouragement of formal value management processes.
- The inclusion of clear provisions for risk management, including open acknowledgement of risk and the proactive use of early warning systems.
- Provisions for non-adversarial problem solving through a clear series of alternative dispute resolution processes.

The PCC contract has undergone a series of amendments since its initial publication, with amendments in both 2008 and again in 2013. Whilst the 2013 edition includes changes aimed primarily at bringing the contract into line with changes in statute provisions, namely amendments to the Housing Grants, Regeneration and Construction Act, other changes instigated in 2008 included more recognition of risk registers, an increased focus on sustainability and the introduction of provisions related to project bank accounts.

## 5.9 Summary

Managing the procurement process is a difficult undertaking. The variety of tasks that are required to be completed and numerous processes to be followed, frameworks to be adopted, standards to be achieved, legislation to abide by, time frames to complete within, budgets that cannot be exceeded, and quality checks that must be adhered to are an ever-changing ocean that must be navigated. In the case of built asset management procurement, this navigation will often fall into the remit of the building maintenance professional. This chapter serves to give a brief overview of just some of the decisions associated with the development of a procurement policy and contract strategy for both responsive (reactive) maintenance and more detailed capital planned maintenance and refurbishment or rehabilitation projects. This chapter attempted to move the focus away from procurement being a simple matter of identifying a procurement route, often dictated by a contract and all too easily placed in a box or silo. Rather, the chapter asks the reader to adopt a businesslike perspective and see procurement as a strategic process that is unique to the situation the maintenance professional faces at the time decisions are made.

The importance of adopting such a proactive approach to procurement cannot be underestimated, nor can the benefits that a proactive and strategic management approach to procurement can bring to the organisation and client. The advantages of early strategic analysis of the business drivers for the project or actions being procured allow the built asset management professional to ensure appropriate client briefings are taken, resulting in a procurement policy that is closely linked to the project business objectives which can ultimately be distilled into a series of critical success criteria mapped to the iron triangle of project management success – time, cost and quality. No longer can procurement be seen merely a choice of route, but rather as a strategic series of unconstrained decisions focused on procurement, pricing strategy and ultimately supported by a suitable contract strategy. Finally, the chapter reveals that despite procurement decisions, contractual arrangements and tendering procedures presenting stand-alone decisions, these processes are interrelated and may be linked in almost any combination, often combinations that cannot be simply fitted into a particular silo or box. The combination selected must ultimately be the one that best satisfies the client's business need.

## References

Atkinson, I. and Wales, S. (2017). Recapping the JCT 2016 changes. https://www. womblebonddickinson.com/uk/insights/articles-and-briefings/recapping-jct-2016-changes (accessed 29 May 2021).

BCIS (2021). Economic significance of maintenance 202. https://service-bcis-co-uk.salford. idm.oclc.org/BCISOnline/Briefings/EconomicBackground/3347?returnUrl=%2FBCISOnli ne%2FBriefing&returnText=Go%20back%20to%20briefing%20summary&sourcePage=Help (accessed 29 May 2021).

Boudjabeur, S. (1997). Design and build defined. In: *13th Annual ARCOM Conference* (ed. P. Stephenson), (15–17 September 1997), King's College, Cambridge. Association of Researchers in Construction Management, Vol. 1, 72–82.

British Standards Institution (2000). *BS ISO 15686–1 Buildings and Constructed Assets – Service Life Planning – Part 1: General Principles*. London: British Standards Institute.

British Standards Institution (2011). *BS8534:2011 Construction Procurement Policies, Strategies and Procedures – Code of Practice*. London: British Standards Institute.

British Standards Institution (2020). *BS ISO 10845-1: 2020 Construction Procurement Part 1: Processes, Methods and Procedures*, 2e. London: British Standards Institute.

Cabinet Office (2011). Government construction strategy. https://assets.publishing.service. gov.uk/government/uploads/system/uploads/attachment_data/file/61152/Government-Construction-Strategy_0.pdf (accessed 29 May 2021).

College of Estate Management (2014). P10464 V1-0 standard forms of contract. Reading: College of Estate Management.

Designing Buildings Wiki (2020). Partnering in construction. https://www. designingbuildings.co.uk/wiki/Partnering_in_construction (accessed 29 May 2021).

Fenwick Elliot (2015). Understanding your design duty – "reasonable skill and care" vs. "fitness for purpose" – mutually incompatible or comfortably coexistent? https://www. fenwickelliott.com/research-insight/annual-review/2014/understanding-design-duty (accessed 29 May 2021).

Glenigan (2015). KPI zone. UK Industry Performance Report. https://www.glenigan.com/ kpi-zone (accessed 29 May 2021).

Higham, A., Bridge, C. and Farrell, P. (2016). *Project Finance for Construction*. Oxon: Routledge.

HM Government (2018). Construction sector deal. https://assets.publishing.service.gov.uk/ government/uploads/system/uploads/attachment_data/file/731871/construction-sector-deal-print-single.pdf (accessed 29 May 2021).

Hughes, W., Champion, R. and Murdoch, J. (2015). *Construction Contracts: Law and Management, 5th Edition*. Oxon: Routledge.

Joint Contracts Tribunal (JCT) (2017). Deciding on the appropriate JCT Contract 2016. https://www.jctltd.co.uk/docs/Deciding-on-the-appropriate-JCT-contract-2016.pdf (accessed 29 May 2021).

Knox, K. (2011). UK: other changes in the JCT 2011 suite of contracts. https://www.mondaq. com/uk/contracts-and-commercial-law/151310/other-changes-in-the-jct-2011-suite-of-contracts (accessed 29 May 2021).

Manchester City Council (2019). Management contractor appointed for Manchester's our town hall project. https://secure.manchester.gov.uk/news/article/8102/management_ contractor_appointed_for_manchesters_our_town_hall_project (accessed 29 May 2021).

Mansfield, J.R. (2008). The use of formalised risk management approaches by UK design consultants in conservation refurbishment. *Engineering Construction and Architectural Management* 16 (3): 273–287.

NBS (2012). National construction contracts and law report 2012. https://www.thenbs.com/knowledge/nbs-national-construction-contracts-and-law-survey-2012 (accessed 29 May 2021).

NBS (2013). National construction contracts and law report 2012. https://www.thenbs.com/knowledge/nbs-national-construction-contracts-and-law-survey-2013 (accessed 29 May 2021).

NBS (2015). National construction contracts and law report 2015. https://www.thenbs.com/knowledge/national-construction-contracts-and-law-survey-2015 (accessed 29 May 2021).

NBS (2018). National construction contracts and law report 2018. https://www.thenbs.com/knowledge/national-construction-contracts-and-law-report-2018 (accessed 29 May 2021).

New Engineering Contract (2021). History of NEC. https://www.neccontract.com/About-NEC/History-Of-NEC (accessed 29 May 2021).

New Engineering Contract (NEC) (2017). *Establishing and Procurement and Contract Strategy Volume 1*. London: Thomas Telford Ltd.

Office of Government Commerce (2003). achieving excellence in construction procurement guide: procurement and contract strategies. https://webarchive.nationalarchives.gov.uk/20110802161443/http://www.ogc.gov.uk/documents/CP0066AEGuide6.pdf (accessed 29 may 2021).

RICS (2012). Contracts in use. A survey of building contracts in use during 2010. London: RICS.

RICS (2013). Developing a construction procurement strategy and selecting and appropriate route. http://www.trentglobal.edu.sg/wp-content/uploads/2017/01/Developing-a-construction-procurement-strategy_GN.pdf (accessed 29 May 2021).

Thompson, S. and Price, A. (2010). JCT 2005: a gentle introduction. https://www.lexology.com/library/detail.aspx?g=86165924-d1f1-4ae2-81ad-e778b04533db (accessed 29 May 2021).

# 6

# Financial Management

Capital Costs

## 6.1 Introduction

Often not reviewed in textbooks focused on capital costs of building work, the refurbishment and maintenance marketplace is a major marketplace for construction professionals. The latest BCIS market analysis (BCIS 2021) suggests the market to be worth £66bn. To give this some context, the report highlights that the refurbishment and maintenance market accounts for almost 33% of construction output in the UK, contributing just under 3% to the national GDP and accounting for 1.23% of total consumer spending (BCIS 2021). Furthermore, the market is growing, the Building Cost Information Service (BCIS) report highlights market growth of 18% since 2012 with ongoing year-on-year increases in output expected over the short to medium term.

Within this marketplace, asset managers will approach or commission large maintenance or refurbishment projects in much the same way as they would new build projects; as a result financial management processes will not change much from those already seen in the new build sector. However, the nature of these projects will be different. Capital projects, the focus of this chapter, will be commissioned as either:

1. Planned maintenance interventions – Often based on the outcomes of condition surveys and will be mapped against the overall organisational budget for planned works. Works such as re-roofing, re-cladding, etc. would be typified by this type of intervention; or
2. Refurbishment – Such capital projects are more intensive and have a much wider scope that is often acknowledged. Egbu (1996) suggests the term 'refurbishment' can be taken to depict any larger scale intervention into an existing structure and can thus include renovation, rehabilitation, extension, improvement, conversion, modernisation, fitting out and repair. In essence, anything undertaken on an existing building to permit its reuse for various specified purposes.

Such works will be undertaken for a variety of reasons and, as Babangida et al. (2012) observed, these can include:

- A need to rearrange or organise space for new uses;
- The need to increase the value of the property;

*Introduction to Built Asset Management*, First Edition. Dr Anthony Higham, Dr Jason Challender, and Dr Greg Watts.
© 2022 John Wiley & Sons Ltd. Published 2022 by John Wiley & Sons Ltd.

- The desire to improve the quality of the space to appeal to changing market dynamics and enhance rental income;
- The need to replace degraded finishes and components;
- The desire to improve the aesthetics of the building;
- The need to respond to energy and carbon requirements through sustainable retrofit;
- The need to replace damaged building envelopes; and
- A desire to enhance the quality of the internal environment to enhance staff well-being and increase workflow efficiency and productivity.

Regardless of the rationale for the intervention and whether this is a more routine planned maintenance intervention or a more substantial programme of refurbishment and rehabilitation, the basic principles of financial management will nevertheless apply. The scope of this chapter therefore looks at how the capital costs of such interventions can be managed and controlled from their inception until the point that the construction phase is completed. Whilst the trigger for these interventions will be different, be it a condition survey or a more strategic review of the organisation's assets, it is acknowledged that any capital investment will require a similar level of organisational review and management to that outlined in this chapter.

## 6.2 Project Appraisal and Developing the Business Case

The historic perception of the construction industry has been evolving over the last 20 years. It can largely still be argued the construction industry's main role is to build and refurbish buildings, with the output from the sector being a new or rehabilitated building, road, bridge, railway or something similar. So whether we are demolishing buildings at the end of their life cycle, producing a new building or range of buildings to enhance the current built environment, or undertaking works to improve an existing building, range of buildings or even a full neighbourhood through refurbishment, or a more extensive master planning focus on place making and community enhancement through extensive remodelling and rehabilitation of the existing built environment; the core view, that construction is, in its simplest, a production process that meets the needs of various customers, cannot be displaced.

Despite this common view, over the last two decades a number of people and organisations have tried to amend the view of the traditional role of the industry. At the forefront of this vanguard of change is Constructing Excellence, an organisation that seeks to drive the change agenda within the construction sector via improvements in performance. In 2009, Constructing Excellence commissioned Andrew Wolstenholme to review the sector and its performance. As part of that review, Wolstenholme suggested that the sector, including government and client organisations, needed to adopt a more business-focused mentality and look at the *built environment* as opposed to the *construction industry* (Wolstenholme 2009). In doing this, client organisations are encouraged not to see a building as a physical artefact or the output of a production process. Rather, they should see buildings as working assets (Constructing Excellence 2006), whereby value is considered from a business perspective, namely minimising building occupation costs whilst aiming to maximise the efficiency of the staff working within it (Constructing Excellence 2006), an argument that could be expanded to suggest the utility or usefulness of the asset to all stakeholder groups should be maximised.

Taking this business perspective forward into the financial decision environment associated with facilities management, it becomes clear that the professional team, whether in-house or external to the client, will need to guide the client through several fundamental decisions. These decisions, depicted visually in Figure 6.1, are the key strategic decisions that will ultimately determine the success or otherwise of their project.

As Figure 6.1 illustrates, prior to commissioning professionals to form the design team, the business need for a project must have first been identified and considered at a strategic level within the organisation. This will include the development of a clear needs-based assessment to ascertain the drivers to moving forward with the development and to appraise the fit in terms of satisfying business objectives and alignment with the strategic direction of their business. Guidance from HM Treasury (2018) recommends that a compelling case for intervention is made. This will include a clear understanding of business needs, the potential benefits, risks, constraints and dependencies associated with the proposal along with a demonstration of the project's holistic fit and synergy with the business's overall strategic direction. According to the Association of Project Management (2021), the typical business case should include the following elements:

- *Strategic Context* – Outline the strategic context and provide a compelling case for change.
- *Economic Analysis* – Provide an economic analysis of the return on investment, based on some level of investment appraisal for the options outlined.

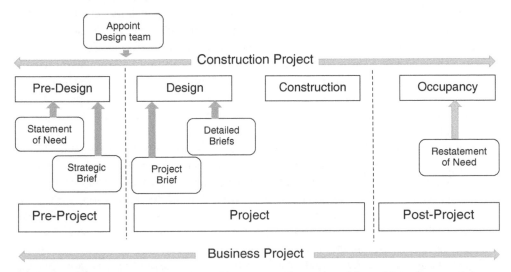

**Figure 6.1** Aligning a business project with a construction project. Adapted from Constructing Excellence 2006.

- *Commercial Approach* – Align with the project's procurement policy and strategy (see Chapter 5 for further discussion of this).
- *Financial Case* – Ascertain the affordability of the proposals to the organisation with an assessment of the likely benefits against costs for each option.
- *Management Approach* – Outline the project governance structure, roles of key parties, life cycle choices and so on.

Within the financial case, a project feasibility appraisal is likely to be required. This provides the business decision makers the opportunity to assess and question the potential options identified before resources are committed and external consultants appointed to develop the construction project. From a purely financial perspective, it is the project appraisal that will ultimately decide not only financially viability, but also the future direction of the project given the clients overall budget, project objectives and alignment with the wider organisational outlook.

### 6.2.1   Optional Appraisal

Forming a key part of the financial and economic case for investment, the options appraisal is undertaken as part of the business case for the proposed project. Mapping this to the construction process, this can be aligned with the Royal Institute of British Architects (RIBA) stage 0 strategic definition, although from an asset management perspective this is likely to be undertaken in-house and before the construction project is conceived, so may well occur before RIBA stage 0 is achieved. However, for smaller organisations, such as one-off clients, this will often mark the first external consultant appointments. During this stage, the RIBA plan of work suggests 'the project will be strategically appraised and defined before a detailed brief is created' (RIBA 2020). At this point, it is highly like an option appraisal will be undertaken, this is likely to require 'a review of a number of sites or alternative options such as extensions, refurbishment or new build' (RIBA 2020). The key outcomes from an option appraisal include a full appraisal of the likely costs of the development and some consideration of potential risks. From a financial management perspective, this review will involve a detailed financial evaluation of the costs associated with the options identified and will typically include the production of an order of cost estimate, which, in essence, is an assessment of each option's affordability and the establishment of a realistic cost limit that considers the potential financial impact of associated risk.

The option appraisal will typically be applied at a project level and will consider the viability of developing a building type potentially considering different scales of development. Alternatively, the options may include an appraisal of the viability of varying levels of rehabilitation of refurbishment of existing assets, alongside consideration of vertical or horizontal extensions to existing buildings within the organisation's asset portfolio. The options will be appraised against how these align with the client's project objectives, not only in terms of their alignment with the client's overall budget, but also in terms of the likely benefits each option would present to the client measured against non-financial project objectives.

## 6.3   Order of Cost Estimate

In terms of the financial management of projects, realistic and reliable cost and time targets will be crucial as the project develops. The latest Royal Institution of Chartered Surveyors (RICS) guidance note advises that 'realistic and reliable cost and time targets are crucial at the outset when making any investment decision' (RICS 2020). To ensure this service is provided, best practice for financial management at this stage of the project is outlined in the *New Rules of Measurement* (NRM1) (RICS 2013). This document advises that at the earliest stage of the project the professional team should develop an order of cost estimate for each possible outcome to outline the likely costs involved.

In this context, the order of cost estimate represents the first attempt to estimate the cost of the proposed building. Although, traditionally, this view of initial costs would have been provided as a point cost or a clearly defined cost in the form of a single number, this is now discouraged, with professionals often seeking to budget range with the confidence intervals surrounding the prediction outlined to better communicate the risks involved with cost forecasting at this early stage in the project's development (RICS 2020). Despite many presenting this initial estimate using a budget range rather than a precise figure, the *New Rules of Measurement: Order of Cost Estimating and Cost Planning for Capital Building Works* produced by the Royal Institution of Chartered Surveyors (RICS 2013), asserts that the order of cost estimate is produced to both define the cost limit for the project whilst also evaluating whether or not the proposed project is feasible. The constituent elements of the order of cost estimate are shown in Table 6.1.

### 6.3.1   Developing the Order of Cost Estimate

As Figure 6.3 illustrates, at this stage in the project the team is likely to be working with less than 10% of the information available that would be needed to complete the capital work. Due to this lack of information and consequent lack of certainty, the accuracy of estimates produced at the feasibility or business case stage of the project tend to be highly variable. In an ideal situation, the client team would have the full range of information identified in the NRM (RICS 2013, p. 20) available and this would include:

- Drawings of the floor and roof plans;
- Elevation details for all the main facades;
- Detailed section drawings through the building;
- Information relating to the storey heights of the building;
- Information relating to MEP (mechanical, electrical and plumbing) installations; and
- Structural design for the building's frame and foundations.

However, this is rarely, if ever, provided given the project remains at a strategic phase and the client will not have triggered the investment required to develop such levels of detail. If it happens that this information is available from previous projects where the client has a standard design scheme they roll out, such as in the house building sector or commercially with fast food restaurants and some supermarket chains, this would facilitate the

**Table 6.1** Order of cost estimate framework (RICS 2013, p. 36).

| Constituent |
| --- |
| **Facilitating works** |
| **Building works estimate** |
| **Contractor preliminaries** |
| **External works (additional allowance)** |
| Subtotal |
| **Main contractor's overheads and profit** |
| Works cost estimate |
| **Project/design team fees** |
| **Other development/project costs** |
| Base cost estimate |
| Risk allowance estimate |
| **Design development risk estimate** |
| **Construction risk estimate** |
| **Employer change risk estimate** |
| **Employer other risk estimate** |
| Cost limit (excluding inflation) |
| **Tender inflation estimate** |
| Cost limit (inc. tender inflation) |
| **Construction inflation estimate** |
| Cost limit (inc. construction inflation) |
| **VAT assessment** |

development of a detailed cost predication. This would likely be developed by a quantity surveyor, who, given the extent of information available would be able to take off elemental quantities for key aspects of the building. This information, alongside reasonably detailed specification information, would allow a reasonably accurate initial forecast of the likely expenditure on the project to be produced, whilst also establishing the maximum cost of the building, termed the *cost limit*.

As alluded to earlier, this idealist situation is far removed from the realities of cost prediction at the business case stage in the project life cycle. Accordingly, the team face the difficult task of predicting likely expenditure and establishing a cost limit for the project with less than 10% of the information available at tender stage (RICS 2020). However, where such information is not available and therefore elemental level analysis cannot be used, the NRM1 (RICS 2013) makes provision for replacing elemental unit quantities (EUQs) with other less accurate techniques such as either the gross internal floor area of the building or, where this is not established, the use of unit rates based on metrics, such as price per bed or price per car parking space.

In situations such as this, appropriate rates based on a comparative project can be used to form the basis of the order of cost estimate. For this reason, the *New Rules of Measurement*

(RICS 2013, p. 19) directs the quantity surveyor to use one of three techniques for producing the order of cost estimate:

- Elemental Method
- Floor Area Method
- Functional Unit Method

These techniques are presented as a decision hierarchy, informed by the information available at the time of the estimate. As discussed above, the elemental method, which is the preferred approach within the NRM hierarchy, given it results in the most accurate estimate of the project cost, requires the most data from other members of the project team, which is rarely available given the resource input required to generate the levels of information required. As a result, in the majority of projects, the initial feasibility or business case estimate will be developed using either floor area or functional unit methods. Due to the unreliable nature of these techniques, the final outcome will often be reported as a budget range, with the maximum cost in that range defined as the cost limit for the project. The next sections show how these different techniques can be used to prepare cost estimates.

### 6.3.2  Developing the Order of Cost Estimate Using the Functional Unit Method

The functional unit method is, in terms of estimating accuracy, the most unreliable technique available, so this would only be applied to a project at the earliest stages in its evolution, where no other information could be provided to give the team an idea of the size and scope of the structure. In the asset management arena this would be very unlikely to occur, unless the organisation was looking to build additional car parking, for instance. A functional unit is a unit of measurement representing the prime use of the building, or a part of the building if the development proposed has a mixture of intended uses; for example, if the client was looking to develop a mixed use scheme which included some leisure facilities, retail floor space and maybe office space at the higher levels.

Using the example of the car park mentioned earlier, the functional unit method can be used to establish a rudimentary cost for the project. For this, however, the team would first need to identify:

- The prime unit of analysis – In the case of a car park this would be *cost per parking space*.
- The number of the prime unit – How many parking spaces will the structure provide; for this example, let's say 230.
- The nature of the car park – Assumed multi-storey.
- Historical cost data relating to prime unit – This could include:
  ○ Commercially available price books and rates databases
  ○ The quantity surveyor's own cost data library
  ○ The Building Cost Information Service (BCIS) database.

To develop the order of cost estimate for the car park using the functional unit method, a range of commercially available functional unit rates are included in Table 6.2. The costs have been adjusted so they reflect tenders in Northern England, and a tender date of 1st Quarter 2023 has been assumed.

**Table 6.2** Functional unit data.

| Building Function | £/functional unit | | | | | | |
|---|---|---|---|---|---|---|---|
| New Build Multi-storey Car Park | Mean | Lowest | Lower Quartile | Median | Upper Quartile | Highest | Sample Size |
| No. of Vehicle Spaces | 17,417 | 8,618 | 10,335 | 16,100 | 23,303 | 30,582 | 9 |

As can be seen from Table 6.2, the range included is wide with functional unit costs from £8,618 to £30,582. Although it must be remembered that this data will potentially include the full array of design and specification options including pre-cast concrete, in-situ concrete and steel frames with various levels of cladding to the exterior and fitout of communal areas. The quantity surveyor must analyse the overall data to ensure they position the costing of the proposed project correctly in this range. This will invariably require more information from the client relating to the anticipated quality and complexity of the proposed car park. In this example, it is assumed the client has narrowed the project scope only slightly to specify that the car park will be a budget-level facility. As a result, the quantity surveyor will need to develop a range of cost. Identifying the minimum and maximum. These are calculated as follows:

- Number of spaces × minimum cost for the car park (in this case the lowest rate is used). So: £8,618 × 230 = £1,982,140
- Number of rooms × maximum cost for the car park (in this case the lower quartile is used). So: £10,335 × 230 = £2,377,050

To fully determine the cost limit for the project, the quantity surveyor will need to use the data to develop a full order of cost estimate as shown in Table 6.1. The development of the full order of cost estimate for the scheme is shown in Table 6.3. From the analysis you can see that the upper cost limit for the car park has been estimated at £5,178,850. Obviously as the quantity surveyor receives more information from both the design team and the client, this estimate would be refined, and the cost limit would be reduced to reflect the increased information and the associated increase in the accuracy of the estimate.

### 6.3.3 Developing the Order of Cost Estimate Using the Floor Area Method

The floor area method, also known as the GIFA (gross internal floor area) rate method, is the most popular tool used by quantity surveyors to develop the feasibility budget and determine the project cost limit. At this stage is highly probable that the designer will have ascertained a rough layout for the building, which allows the gross internal floor area of the proposed building to be established. The GIFA of the building is defined in the *RICS Code of Measuring Practice* (6th edition) as 'the area of a building measured to the internal face of the perimeter walls at each floor' (RICS 2018). At the time of writing, the *Code of Measuring Practice* is being phased out and replaced with the *International*

**Table 6.3** Order of cost estimate for car park based on functional unit data.

| Constituent | % | Minimum (£) | Maximum (£) |
|---|---|---|---|
| **Facilitating works** | | 0 | 0 |
| **Building works estimate** | | 1,982,140.00 | 2,377,050 |
| **Contractor preliminaries** | Included | 0 | 0 |
| **External works (additional allowance)** | Included | 0 | 0 |
| Subtotal | | **1,982,140** | **2,377,050** |
| **Main contractor's overheads and profit** | Included | 0 | 0 |
| Works cost estimate | | **1,982,140** | **2,377,050** |
| **Project/design team fees** | 10% | 198,214 | 237,705 |
| **Other development/project costs** | | 0 | 0 |
| Base cost estimate | | **2,180,354** | **2,614,755** |
| Risk allowance estimate | | | |
| **Design development risk estimate** | 7% | 152,625 | 183,033 |
| **Construction risk estimate** | 7% | 152,625 | 183,033 |
| **Employer change risk estimate** | 3% | 65,411 | 78,443 |
| **Employer other risk estimate** | 3% | 65,411 | 78,443 |
| Cost limit (excluding inflation) | | **2,616,425** | **3,137,706** |
| **Tender inflation estimate** | 1.34% | 35,060 | 42,045 |
| Cost limit (inc. tender inflation) | | 2,651,485 | 3,179,751 |
| **Construction inflation estimate** | 8.76% | 232,270 | 278,546 |
| Cost limit (inc. construction inflation) | | **2,883,755** | **3,458,297** |
| **VAT assessment** | | Excluded | Excluded |

*Property Measurement Standard.* It is not expected that the definition or use of the GIFA unit will change as part of this process.

At this point, an example more aligned to the topic of this book will be used. Your client is a major hotel chain looking to convert a disused office building into a new budget hotel. Given that the floor area of the building is available from the sales particulars and initial surveys of the building, it is possible to apply the floor area method to establish a business case estimate for this project. To inform this estimate the team would first need to identify:

- The extent of the refurbishment – extensive, strip back the building to core frame and redevelop.
- The anticipated floor area – 1800 m$^2$.
- Historical cost data relating to floor area costs – this could again include:
  ○ Commercially available price books and rates databases
  ○ The quantity surveyor's own cost data library
  ○ The Building Cost Information Service (BCIS) database.

To develop the order of cost estimate for the hotel refurbishment project using the floor area method, a range of $£/m^2$ GIFA rates have been sourced; these are included in Table 6.4. The costs have been adjusted so they reflect tenders in Northern England, and a tender date of 3rd Quarter 2022 has again been assumed.

As can be seen from Table 6.5, the range included is extremely wide – the unit costs range from £1,060 to £5,311 per square metre of gross internal floor area (GIFA). However, the construction cost data available is clustered around the lower end of the range with most of the sample resting between £1,100 and £2,500. In terms of positioning the project in the data range available, the level of specification, scale of the project and anticipated specification for the hotel would influence the decision on where to place the project but it is anticipated it will fall within this lower range.

To facilitate the development of the estimate in the example, it has been assumed the client has narrowed the project scope only slightly to specify that the hotel will be a budget-level facility with a 2* rating. Once again, a range of costs will be identified with the range parameters specified as a minimum and maximum price. Calculated as follows:

- Number of rooms × minimum cost for the hotel (in this case the lowest rate is used). So: £1060 × 1800 = £1,908,000
- Number of rooms × maximum cost for the hotel (in this case the lower quartile is used). So: £1535 × 1800 $m^2$ = £2,763,000

Once again, to fully determine the cost limit for the project, the quantity surveyor will need to use the data to develop a full order of cost estimate as shown in Table 6.1. The development of the full order of cost estimate for the scheme is shown in Table 6.5. Although the range at this point in time is wide, this will reduce as more certainty is given related to the scheme's design and specification. This will invariably need to the inclusion of less risk premium in the project costings. In this hypothetical example, risk allocations have been assumed. However, in practice, such risk allocations would be based on a detailed analysis of the risks associated with the specific project; even at this early stage in the project life cycle high-level strategic risks can be identified and analysed. For a more comprehensive discussion of this please review Chapter 9.

In some situations, the order of cost estimate or the business case estimate can be provided with a greater level of certainty. This is typically where the professional team or client have access to their own historical cost data, and this can be reasonably applied to the project under consideration.

**Table 6.4  £/m2 GIFA data for conversion and rehabilitation.**

| Building Function | $£/m^2$ gross internal floor area | | | | | | |
|---|---|---|---|---|---|---|---|
| New Build Hotels | Mean | Lowest | Lower Quartile | Median | Upper Quartile | Highest | Sample Size |
| Hotels | 2057 | 1060 | 1,535 | 1,752 | 2054 | 5311 | 10 |

**Table 6.5   Order of cost estimate for hotel based on floor area (£/m² ) data.**

| Constituent | % | Minimum | Maximum |
|---|---|---|---|
| **Facilitating works** | | 0 | 0 |
| **Building works estimate** | | 1,908,000 | 2,763,000 |
| **Contractor preliminaries** | Included | 0 | 0 |
| **External works (additional allowance)** | Included | 0 | 0 |
| Subtotal | | **1,908,000** | **2,763,000** |
| **Main contractor's overheads and profit** | Included | 0 | 0 |
| Works cost estimate | | **1,908,000** | **2,763,000** |
| **Project/design team fees** | 10% | 190,800 | 276,300 |
| **Other development/project costs** | | 0 | 0 |
| Base cost estimate | | **2,098,800** | **3,039,300** |
| Risk allowance estimate | | | |
| **Design development risk estimate** | 10% | 209,880 | 303,930 |
| **Construction risk estimate** | 15% | 314,820 | 455,895 |
| **Employer change risk estimate** | 3% | 62,964 | 91,179 |
| **Employer other risk estimate** | 3% | 62,964 | 91,179 |
| Cost limit (excluding inflation) | | **2,749,428** | **3,981,483** |
| **Tender inflation estimate** | 1.34% | 36,842 | 53,352 |
| Cost limit (inc. tender inflation) | | **2,786,270** | **4,034,835** |
| **Construction inflation estimate** | 8.76% | 244,077 | 353,452 |
| Cost limit (inc. construction inflation) | | **3,030,348** | **4,388,286** |
| **VAT assessment** | | Excluded | Excluded |

Consider the case of a local authority who has undertaken a programme of works or been investing in its school estate over a period of several years; they would have access to much more accurate and reliable cost data from those previous projects with which to estimate the likely cost of the current project. As can be seen from the data in Table 6.6, the data from the construction of a series of new primary schools has been used to predict the costs of extending a fifth; this data is based not on tender outcomes but on final account information to give the client a much more accurate view of likely costs (shown in Table 6.7) once risks have been fully realised.

This data can then be used to inform the order of cost estimate for the fifth school in the series. In this example, there is no need for a minimum and maximum column given the client's use of mean cost data and their knowledge of these projects.

Whilst this stage in the cost modelling covers the extension, the refurbishment of the existing structure is not costed. This cannot be modelled reliably using the same methods, given so little information was available that any level of reliable cost prediction was not possible.

### 6.3.4 Developing the Order of Cost Estimate Using the Elemental Method

The elemental method represents the most accurate approach to producing the initial feasibility estimate for a project. This fact is fully recognised and documented in the *New Rules of Measurement*, which stresses the importance of using elemental analysis as the most accurate and therefore the preferred approach to estimating the cost of a proposed project. However, the method is also the most demanding in terms of design and specification development, but the complexity of refurbishment works makes the elemental cost

**Table 6.6** Client historic cost data for schools within the local area.

| Name | School A | School B | School C | School D | Average |
|---|---|---|---|---|---|
| Building Total | 1,203.81 | 1,512.74 | 1,763.30 | 1,473.54 | 1,609.88 |
| Prelims | 157.40 | 138.97 | 490.58 | 472.05 | 314.75 |
| Total | 1,361.20 | 1651.71 | 2753.88 | 1945.59 | 1924.63 |

**Table 6.7** Order of cost estimate for 900 $m^2$ school extension based on floor area ($£/m^2$) data.

| Constituent | % | Maximum |
|---|---|---|
| **Facilitating works** | | 0 |
| **Building works estimate** | | 1,448,892 |
| **Contractor preliminaries** | | 283,275 |
| **External works (additional allowance)** | Excluded | 0 |
| Subtotal | | **1,732,167** |
| **Main contractor's overheads and profit** | Included | |
| Works cost estimate | | **1,732,167** |
| **Project/design team fees** | 10% | 173,217 |
| **Other development/project costs** | Excluded | 0 |
| Base cost estimate | | **1,905,384** |
| Risk allowance estimate | | |
| **Design development risk estimate** | 5% | 95,269 |
| **Construction risk estimate** | 10% | 190,538 |
| **Employer change risk estimate** | 3% | 57,162 |
| **Employer other risk estimate** | 3% | 57,162 |
| Cost limit (excluding inflation) | | **2,305,514** |
| **Tender inflation estimate** | 0% | |
| Cost limit (inc. tender inflation) | | **2,305,514** |
| **Construction inflation estimate** | 0% | |
| Cost limit (inc. construction inflation) | | **2,305,514** |
| **VAT assessment** | | Excluded |

method the only sensible way to cost such projects. Although a comprehensive explanation of the elemental cost planning process is provided later in the chapter, at this stage, it is nevertheless worth considering how this technique can be used to develop initial cost budget cost estimates at the business case stage.

Consider, once again, the example of the school above. It was not possible to use the floor area method to cost the refurbishment aspect of the project given the general level of complexity associated with refurbishment works. However, given that the client had instructed a detailed stock condition survey of their school estate, which ranked most buildings as amber (in need of refurbishment in the medium term) or red (short-term action needed), it has been possible to use this data to estimate, although with high levels of uncertainty until future investigations are undertaken, a cost limit for the project. The costing for the refurbishment of the existing structure is depicted in Table 6.8. It can be seen that the rough scope of the refurbishment has been ascertained for the existing structure and a rate devised from previous comparable refurbishment projects undertaken by the local authority.

**Table 6.8** Order of cost estimate refurbishment of the existing school building.

| Refurbishment Works | Qty | Unit | Rate | £ |
|---|---|---|---|---|
| **Existing building condition works** | | | | |
| Remove existing roof and alteration work | | Item | | 8,000.00 |
| Connect new building to existing | | item | | 40,000.00 |
| New roof to existing building to line in with | | | | |
| new build extension | 96 | m$^2$ | 184 | 17,627.90 |
| Internal alterations to existing school | 373 | m$^2$ | 536 | 199,767.61 |
| Essential mechanical works | | Item | | 14,000.00 |
| Upgrading existing boilers | | item | | 75,000.00 |
| Repairs to structural condition of existing building | | Item | | 8,000.00 |
| DDA | | Item | | 8,000.00 |
| New disabled toilet | | item | | 8,000.00 |
| New intruder alarm to existing school | | item | | 15,000.00 |
| Alterations telephone system | | item | | 4,000.00 |
| **New build provisional sums** | | | | |
| Planting / play equipment | | Item | | 50,000.00 |
| Removal of temporary buildings | | Item | | 30,000.00 |
| Artificial playing pitch/fenced | | Item | | 60,000.00 |
| Fencing (repairs) | | Item | | n/a |
| New lift | | item | | incd in ext |

*(Continued)*

**Table 6.8** (*Continued*)

| Refurbishment Works | Qty | Unit | Rate | £ |
|---|---|---|---|---|
| **Incoming services** | | | | |
| New increased mains cold water supply (existing lead service) | | Item | | 23,000.00 |
| New increased gas supply | | item | | 30,000.00 |
| New increased electric supply | | item | | 25,000.00 |
| New telephone cable alterations | | item | | 7,000.00 |
| | | | | |
| **External works** | | | | |
| Remove trees – large | 2 | nr | 842 | 1,683.22 |
| Remove trees – small | 2 | nr | 612 | 1,224.16 |
| Repointing, brick coping and making good to | | | | |
| existing low-level wall | 36 | m | 153 | 5,508.72 |
| Planting to habitat area | 450 | m$^2$ | 38 | 17,214.75 |
| Nursery play area and fence | 400 | m$^2$ | 92 | 36,724.80 |
| Landscaping to front of building | 136 | m$^2$ | 38 | 5,202.68 |
| Tarmac and sub-base to hard play areas | 2,100 | m$^2$ | 77 | 160,671.00 |
| New covered area for bicycles | | Item | | 14,000.00 |
| Sundry external painting works | | Item | | 8,000.00 |
| Remedial works to perimeter boundary line | | Item | | 7,000.00 |
| | | | | |
| **General additional works** | | | | |
| Reline existing car park | | item | | 3,000.00 |
| New crossover and gates | | item | | 15,500.00 |
| Temporary classroom | | item | | 50,000.00 |
| Asbestos removals (mobile) | | item | | 8,000.00 |
| AVA costs for ICT equipment / termination | | item | | 30,000.00 |
| Whiteboard costs (Promethean) | | item | | 3,000.00 |
| Fitted furniture in new build | | item | | incd in blg rate |
| Pre-fab garage on base | | item | | 5,000.00 |
| External signs | | item | | 3,000.00 |
| Move costs / overtime | | item | | 7,500.00 |
| | | | | **1,004,624.84** |

## 6.4  Cost Planning

Once the project has moved beyond the feasibility or business case stage in its development, the employer has committed to developing the scheme to the end of the pre-contract phase. In other words, a full consultant team will have been appointed and the scheme will move forwards through the design stages ready to be tendered in the marketplace. As the design team starts to develop the conceptual scheme approved by the client at the project appraisal stage, costs will continue to be managed, with the project's quantity surveyor providing regular reports to the client addressing cost prediction, control and financial risk management issues. Financial reports will typically be associated with the production of increasingly detailed estimates for the overall project costs. These will be aligned to the key employer reporting stages, which are typically aligned with phase of design development, whilst providing a cost control and an oversight role; in other words, ensuring the client's budget is adhered to and making sure the project does not overshoot in financial terms the predetermined cost limit approved by the client senior team at the pre-project stage.

To ensure the project remains within the scope of the cost limit outlined in the business case, a financial management technique known as 'design to cost' will be adopted. Often associated with production, design to cost is a management tool designed to track the project through its design phase within the central aim of ensuring it does not exceed the budget allocated for its development. Design to cost therefore acts as a benchmark to see how far the development of a project or, in this case, the building has progressed whilst also ensuring the design remains financially balanced. The technique further allows the design team to pre-empt potential problems by identifying areas where the project is likely to exceed the employer's available budget, so the design team can identify creative solutions to overcome these difficulties. Implemented into construction, the design to cost management technique is widely known as *cost planning*. Although some textbooks will argue cost planning is in reality only one aspect of a much more complex pre-contract cost control system, the accepted industry best practice guidance, NRM1, defines this process simply as cost planning; for that reason this book will adopt the same terminology.

The cost plan is defined in the *New Rules of Measurement* as:

> The critical breakdown of the cost limit for the building into cost targets for each element of the building. It provides a statement of how the design team proposes to distribute the available budget amongst the elements of the building, and a frame of reference from which to develop the design and maintain cost control. It also provides both a work breakdown structure and a cost breakdown structure which by codifying can be used to redistribute work in elements to construction works packages for the purpose of procurement (RICS 2013, p. 12).

From the definition provided in the NRM it is clear that the cost plan forms the principal tool for pre-contract financial management of the construction project. Not only does the cost plan provide a breakdown of the overall project budget into core cost centres referred to as elements, it also provides a frame of reference against which the design team can develop the scheme, whilst allowing the opportunity to ensure costs are fully managed and controlled,

**Table 6.9** Constituent elements of a cost plan (RICS 2013, p. 37).

| Constituent |
| --- |
| **Building works estimate** |
| **Main contractor's preliminaries estimate** |
| **Subtotal** |
| **Main contractor's overheads and profit estimate** |
| Works cost estimate |
| Project/design team fee estimate |
| **Consultant's fees** |
| **Main contractor's pre-construction fee estimate** |
| **Main contractor's design fee estimate** |
| **Subtotal** |
| Other development/project cost estimate |
| Base cost estimate |
| Risk allowances estimate |
| **Design development risk estimate** |
| **Construction risk estimate** |
| **Employer change risk estimate** |
| **Employer other risk estimate** |
| Cost limit (excluding inflation) |
| Tender inflation estimate |
| **Cost limit (inc. tender inflation)** |
| Construction inflation estimate |
| **Cost limit (inc. construction inflation)** |
| **VAT assessment** |

both against the overall budget, as broken down in Table 6.9, and against individual cost centres that form part of the 'building works estimate', which will be discussed later.

Cost planning, however, must not be seen as a single activity; that will only happen at the commencement of the design phase. To liken cost planning to photography, the cost plan may appear to many to be like a photograph, a view of the project's financial position captured in a single point of time. In reality, cost planning is a continuous process that spans the full design phase of the project. So, returning to the photographic analogy, cost planning can be likened to a video. The cost plan is an evolving document consisting of numerous snapshots of the project that are linked together to provide a continuous image; in this case, a real-time view of cost mapped against the employer's available budget. The fundamental objectives of the cost plan are defined by the NRM1 (RICS 2013, p. 50) as:

- To ensure the employer received value for money;
- To make both employers and designers aware of the cost consequences of their design or proposal;

- To provide advice to designers that enables them to arrive as a practical and balanced design within the scope of the budget;
- To keep project expenditure within the cost limit approved by the employer;
- Provide robust cost information upon which the employer can make informed decisions.

To make this process as useful for the employer, the cost plan is captured and reported at key project milestones. These are aligned to the RIBA plan of work, as shown in Figure 6.2, and constitute the formal cost plan requirements of the NRM1, although the exact reporting requirements may be amended by the employer.

As shown in Figure 6.2, NRM1 identifies three 'formal cost plans' defined by the NRM1 as 'the elemental cost plan which is reported to the client on completion of a specific RIBA work stage' (RICS 2013, p. 13). These cost plans represent formal reporting stages during design development, whereby the design team is required to provide a full project cost report to the client. Whilst each formal cost plan represents a progression of order of cost estimate, the level of detail and certainty enhances at each reporting stage. With design information evolving, a clearer picture of the project will emerge at each cost planning stage. As a result, the reports the client receives or the reports the team presents to the organisation's senior leadership team become increasingly more comprehensive, presenting an increasingly more accurate view of the project's expenditure profile than the previous. The three formal cost plan reporting stages identified in the NRM1 are:

- *Formal Cost Plan 1* – Prepared when the scope of work is fully defined, and key criteria specified but no detailed design has commenced.
- *Formal Cost Plan 2* – Prepared when design development is complete.
- *Formal Cost Plan 3* – The final cost plan produced, this is based on the completed technical designs, specifications and detailed information now available to the quantity surveyor.

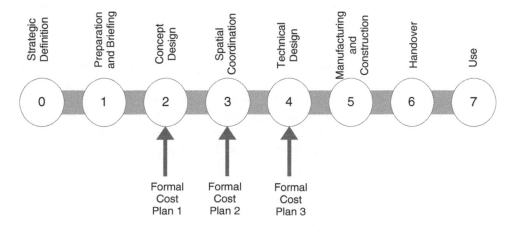

**Figure 6.2    NRM1 Key reporting milestones. Adapted from Higham et al. 2017.**

At each point the client, or the organisation's senior leadership team, will be presented with a detailed report updating them on the financial position of the project and any anticipated financial risk identified by the quantity surveyor as they scan the project's forward horizon.

It is worth of noting at this point that a fourth reporting stage is identified within the NRM: Formal Cost Plan 4 deals with life cycle and maintenance costs into the occupancy phase. For full discussion of this please review Chapter 7.

### 6.4.1 Preparing the Cost Plan

As explained earlier, each of the three formal cost plans represents a progression and refinement of the estimate provided in the order of cost estimates at the outset of the project. For that reason, each cost plan is really just a more detailed, logical progression from the first. With each cost plan, the professional team provides the level of accuracy achieved when compared to the tender and the final account figure should be enhanced, thereby reducing the levels of variability in the estimate. Whilst the NRM refers to four formal reporting stages, the first being the order of cost estimate (OCE), the latest RICS professional statement offers a six-stage reporting model, with each stage representing a reporting step to the client. It is proposed that this sequential, stepped approach will 'provide a framework for cost prediction consistency', ensuring the client is clear about the levels of granularity within the cost prediction process. Whilst for some this will introduce confusion with the formal reporting stages in NRM, as Figure 6.3 demonstrates, the two can be interwoven.

The improvements in accuracy noted in Figure 6.3 are based on enhanced design development and input maturity (specification and project information development). These improvements enable the team to increase the depth of analysis each iteration of the cost report provides. With this in mind, the NRM sets out a hierarchy of methods that can be used when preparing a formal cost plan. These include:

1. The elemental method based on elemental unit quantities.
2. The elemental method based on approximate quantities.
3. If neither of the above are possible due to lack of design information, then the elemental method based on cost/m$^2$ GIFA should be used.

Progression from the initial OCE will ideally be via the elemental method based on elemental unit quantities, and onwards towards a very detailed cost plan based on elemental approximate quantities, assuming the project is procured based on a bill of quantities. However, in reality, this is not always possible as most projects will be procured using either a schedule of rates or a design and build approach. For this reason, it is more realistic to expect the team to apply all three techniques during the pre-contract financial management process, depending on the point at which procurement occurs:

- *Formal Cost Plan 1* – Typically produced based on Cost/m$^2$ GIFA.
- *Formal Cost Plan 2* – Although the level of available information will have improved, it is unlikely this will be sufficient for the quantity surveyor to develop an estimate based on elemental unit quantities.

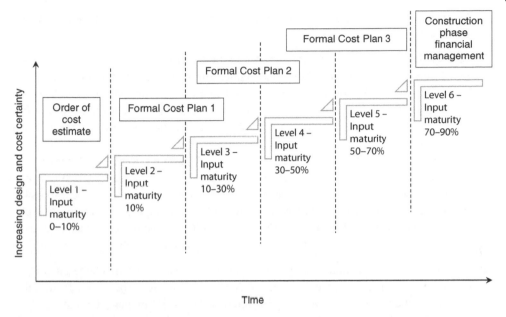

**Figure 6.3** Financial report milestones and NRM formal cost plans. (Adapted from RICS 2013, 2020).

- *Formal Cost Plan 3* –The levels of information now available should mean that the quantity surveyor will have been able to provide a highly detailed cost plan based on elemental approximate quantities.

### Defining Group Elements, Elements, Sub-Elements and Components

In addition to detailing the main approaches to the development of the cost plan, NRM1 part four provides a breakdown of the elements that make up a project. As explained earlier, this breakdown of the building works estimate into elements allows the design team and employer to ensure costs are comprehensively controlled.

However, given the varying approaches to estimating identified earlier, NRM1 also provides five principal levels of detail that can be incorporated into the cost plan. These levels of detail are outlined comprehensively in the fourth section of NRM1 called the 'tabulated rules of measurement' (RICS 2013, pp. 75–327). The tabulated rules of measurement set out the different levels of description and measurement that the quantity surveyor should use. These levels of measurement are defined below:

- *Group Element* – The main headings used to describe the structure of the cost plan; examples of group elements include substructure, superstructure, internal finishes, etc.
- *Element* – A major part of the building that is applicable to a variety of building types; for example, all buildings regardless of their use will have external walls.
- *Sub-Element* – Part of an element; cost targets are set at sub-element level in a similar way to those set for elements.

- *Component* – This is a measurable item that forms part of both an element and sub-element.

To further help you understand the differences between these levels of measurement, Table 6.10 provides an example of the levels of measurement in relation to internal finishings to walls. As noted above, elements and sub-elements are applicable to all building types, so this information could be applied to any scenario from replacing the wall finishes on a three-storey office building refurbishment to finishes for new walls as part of a school extension.

The decision of applicability, in terms of the levels of NRM to be applied, ultimately rests with the surveyor, but this will be largely dependent on both the levels of design information available and the method of production selected. For example, it would be impossible to measure sub-elements without comprehensive technical designs and without using elemental approximate quantities. Consequently, the levels of measurement expected at each cost planning stage would be significantly more detailed and comprehensive; for example:

- *Formal Cost Plan 1* – Would usually be prepared on the basis of group elements or more realistically, elements, measured using $m^2$/GIFA. Using the information from Table 6.11 this would be internal finishes (group element) or wall finishes (element).
- *Formal Cost Plan 2* – Expanding from formal cost plan 1, it would be expected that elemental unit quantities will be applied and the analysis will be broken down into subelements. Again, using the information from Table 6.11 this would be wall finishes.
- *Formal Cost Plan 3* – This would be almost a bill of quantities in terms of measurement precision, so it would be expected that the quantity surveyor will break costs down to component level. Once again using the information from Table 6.11, costs and associated quantities would now be presented for finishes to walls and columns with additional details given, picture and dado rails if used and finally details of any proprietary protections such as those seen in healthcare environments.

Applying this to practice, the quantity surveyor would simply work to the most detailed level possible with the information given. As shown in Table 6.11, the accuracy and detail provided in the estimate develops as design maturity and information availability increase.

**Table 6.10** Tabulated measurement rules (RICS 2013, pp. 143–144).

| Level 1: Group Element | Level 2: Element | Level 3: Sub-Element | Level 4: Component |
|---|---|---|---|
| 3. Internal finishes | 3.1 Wall finishes | 1. Wall finishes | 1. Finishes to walls and columns: details to be stated. |
| | | | 2. Picture rails, dado rails and the like; details to be stated. |
| | | | 3. Proprietary impact and bumper guards, protection strips, corner protectors and the like; details to be stated. |

**Table 6.11** Design maturity mapped to estimating techniques (RICS 2020, p. 27).

|  | Level 1 | Level 2 | Level 3 | Level 4 | Level 5 | Level 6 |
|---|---|---|---|---|---|---|
| Design Maturity | 10% | 10–30% | 30–50% | 50–70% | 70–90% | Varies |
| Type of prediction model | Historical information & judgement | Parametric estimate | Semi-detailed unit costs | Detail (quantity based) | Detailed (quantities and full specification) | Economic valuation |

For the refurbishment of an office building, at the end of concept design (RIBA stage 2), the surveyor is likely to have an idea of the building layout from which the area of wall finishes could be established. It is also likely some specification information will also be available (e.g. fire-rated plasterboard applied using a technique called dry lining). This would suggest design maturity has reached stage 3, allowing the surveyor to apply elemental units to the scheme and develop a cost plan based on semi-detailed unit costs. However, the lack of detailed specification would prevent them applying approximate (detailed) quantities until further information was provided by the design team. The next sections build on this simple example and provide examples showing how different cost planning techniques are applied.

### Data Analysis and Manipulation
The cost data available from various sources such as in-house databases, commercial price books and systems such as BCIS is often presented in a standard format but it is usually also presented at the time it was loaded into the system; as such it does not represent prices at the current time. This is what economists call 'nominal'. Nominal prices are money values that are presented in their original year of capture. To remove this ambiguity, we need to use 'real' monetary values, or values of money that are consistent with each other both in terms of time and location, whilst also ensuring factors such as risk, uncertainty, contingency and inflation are resolved. To remove ambiguity within the data, the use of 'real' monetary values, or values of money that are consistent with each other both in terms of time and location are adopted via adjustment to the data.

Adjustment to cost data is provided in three stages within the cost planning process as illustrated in Figure 6.4. The first adjustment made to the data is baselining. To help the surveyor in this regard, BCIS publishes a series of indices. The key indices for predicting price movements are the Tender Price Index (TPI), which records macro-level change in the economy, and the location index, which records micro-level change between geographic locations. Both indices provide these figures on a quarterly basis. It is also important to note that other commercial organisations such as AECOM provide similar indices for use by quantity surveyors; however, these are not easily interchangeable as the project evolves so it is important to standardise the index used.

The BCIS Tender Price Index (TPI) provides records of historic price movements and predictions of future tender price increases or decreases. Forecasts of future price

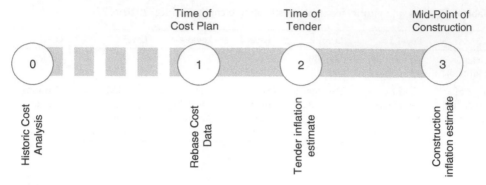

**Figure 6.4** Data adjustment points for cost planning.

movements are made by bringing together a range of key economic data, sourced from government departments such as the Treasury and Department for Business, Innovation and Skills, the Bank of England, and other eminent economic forecasting organisations together with available market intelligence. The BCIS TPI was set at 100 in 1985; this is referred to as the base date.

The index allows the surveyor to then calculate price movement on a quarterly basis using a simple formula given the in NRM (RICS 2013 p. 42):

$$P = \frac{(Index\ 2 - Index\ 1)}{Index\ 1} * 100$$

where

P = percentage addition or subtraction
Index 1 = the index at the base date of the cost data
Index 2 = Index at the current estimate base date.

Adjustment to cost data is provided in three stages within the cost planning process, as illustrated in Figure 6.4. The first adjustment made to the data is baselining. It is likely, when using historic cost data, that the data will be derived from different sources. These data points will need to be adjusted to ensure all the data is comparable and consistent. The application of the above formula allows all data to be initially baselined to the current point in time when the estimate is being produced.

Say, for example, if the project informing the *parametric estimate* was based in Penrith and had a tender base date of June 2004, giving it a TPI index of 215 (2nd Quarter, 2004), it is anticipated the proposed project will have a tender date of May 2021, giving it a TPI index of 329(f). That represents a price movement of 53.02% calculated using the formula thus: P = ((329 − 215) ÷ 215) × 100.

It is important at this stage to explain an important distinction in economics and finance between *price* and *cost*. Although this chapter has used the word *cost* extensively, when we are 'cost planning' we are actually attempting to forecast the likely tender price and ideally the likely final account price for the building under consideration. However, in other areas of practice, such as dealing with fluctuations under a contract, it would be important for the surveyor to consider changes in construction costs, those are the costs incurred by the main contractor at the first tier in the construction supply chain, related to movements in the cost of labour, materials and plant.

For this reason the BCIS service also produces an index known as the *General Building Cost Index (GBCI)*; separate to the tender price index, this index provides evidence of cost movements, so as costs rise, the index number will increase and as costs fall the index number will decrease. Some would expect the GBCI and TPI to be highly positively correlated, whereby an increase in one would cause an increase of near identical proportion in the other, as the contractor passes on the increases in costs to the employer through higher tender prices. However, the highly competitive nature of construction prevents such as correlation. When work is scarce, such as immediately after the market crash in 2008, contractors would seek to absorb increases in cost in an attempt to win work. In some situations, contractors desperate for turnover would often submit a below-cost bid (known as a suicide bid) simply to keep trading, in the hope that lucrative design changes would ease the project to profitability. Equally, when work is more abundant and contractors are managing swelling forward-order books, such as in the early part of the this century (2000–2007), contractors will often seek to price this risk and increase their profit margins and still win work. This can be seen graphically in Figure 6.5, which shows a plot of quarterly price changes over the period from 2000 to

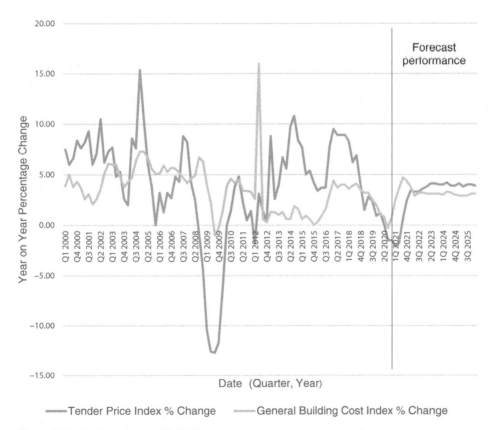

**Figure 6.5** Tender prices and building cost year-on-year percentage change.

2026. It should be noted that as the financial crisis impacted on the construction industry in mid-2008, tender prices dipped considerably. As a result, the period between 2007 and 2010 shows that construction costs are rising whilst tender prices are falling, suggesting the scarcity of work during this period is pressuring contractors to absorb the inflationary pressure on input costs. As a result, margins will have been reduced, and some contractors will have bid below cost simply to remain trading. Yet as the economy begins to recover in the period from 2012 to 2026, tender prices are forecast to increase at a more rapid rate than costs, suggesting that as demand returns, contractors will start to increase their margins and therefore increase profitability. Finally, Figure 6.5 demonstrates the forecasted effect of COVID-19 pandemic on construction demand; although the data is forecast, it does suggest the industry has weathered this economic shock reasonably well when compared to other economic sectors.

In addition to the macro market adjustment, we must also consider the impact of regional variations in economic performance. Location factors can be used within indices published by the BCIS to take to adjust the data for the micro effects of the location where the building is to be constructed or an existing building is to be rehabilitated. Table 6.12 illustrates some of the BCIS location factors for 2021.

As Table 6.13 illustrates, these factors can be reduced to a very specific level of focus; in this example, the city of Leeds in the county of West Yorkshire in the Yorkshire and Humber region of England is examined. The data in Table 6.13 show that Leeds has a location factor of 93 based on a sample of 76 projects.

The final point to make about the adjustment for price movements relates to when these should be included in the cost plan. The order of cost estimate and the three separate formal cost plans should all use cost data adjusted for location and inflation. However, the NRM states that these adjustments should only be made to the base date of the report.

**Table 6.12**  BCIS 2021 location factors.

| Area | Location Factor | 90% Confidence Interval | Sample |
| --- | --- | --- | --- |
| North East | 92 | 92–93 | 476 |
| North West | 97 | 97–98 | 642 |
| Yorkshire and the Humber | 93 | 93–94 | 949 |
| East Midlands | 105 | 104–106 | 1025 |
| West Midlands | 95 | 95–96 | 659 |
| East of England | 101 | 100–102 | 1023 |
| London | 124 | 123–125 | 1035 |
| South East | 108 | 107–108 | 1519 |
| South West | 103 | 102–103 | 878 |
| Wales | 94 | 93–95 | 410 |
| Scotland | 91 | 91–92 | 1303 |
| Northern Ireland | 56 | 55–57 | 203 |
| Islands (Man, Scilly and Channel) | 110 | 108–112 | 148 |

**Table 6.13** BCIS 2021 location factors, hierarchy of levels.

| Overall Area | County | Town or City | Location Factor | 90% Confidence Interval | Sample |
|---|---|---|---|---|---|
| **Yorkshire and the Humber** | | | 93 | 93–94 | 642 |
| | East Riding and North Lincolnshire | | 94 | 91–94 | 143 |
| | North Yorkshire | | 98 | 96–100 | 102 |
| | South Yorkshire | | 94 | 92–95 | 154 |
| | West Yorkshire | | 92 | 91–93 | 243 |
| | | Bradford | 90 | 88–92 | 75 |
| | | Calderdale | 89 | 86–92 | 21 |
| | | Kirklees | 95 | 91–99 | 36 |
| | | **Leeds** | **93** | **91–96** | **76** |
| | | Wakefield | 91 | 89–94 | 35 |

Therefore, as the project progresses the cost data used will require continuous updating, or *rebasing*, as it is termed on BCIS. The quantity surveyor must, however, also add their prediction of both tender and construction inflation. The NRM stipulates that tender inflation is to be taken from the date of the cost plan to the anticipated base date of the tender period (usually the mid-point of the tender period). Construction inflation is to be taken from the tender period to the mid-point (time not expenditure) of the construction project.

### 6.4.2 Formal Cost Plan 1

At this stage in the project's development, it is likely the surveyor will have only received initial, conceptual designs from the architect. It is worth noting that although the project appraisal and OCE will have identified a direction for the project, for example new construction rather than refurbishment, or construction on site A rather than site B, the appearance and layout of the building will not have been evaluated. As a result, the conceptual design stage represents the point in the project when layout and appearance are resolved. It is therefore likely the surveyor will receive multiple designs and will be asked to produce several different cost plans based on these designs. Each alternative design is likely to include plans for each floor, the main elevations and some limited specification information. For that reason, it is unlikely that measurement (quantification) of the building will be possible. Rather, the task of the surveyor is to evaluate the affordability of each design, comparing it to the cost limit determined earlier and to break down the overall cost limit presented in the business case into elemental cost limits.

The scarcity of specification information alongside the lack of detailed measurement over and above simple floor areas, together with limited historic cost data, mean the estimate developed at this stage will be limited in scope. As such, the floor area approach is the simplest and, in some respects, the least resource-intensive technique available for the production of an elemental cost plan. Is normally adopted at this stage to provide what the RICS practice note (RICS 2020) refers to as a *parametric estimate*. Parametric estimates are estimates derived largely from analysis of similar past projects with the aim of giving some indication of financial distribution across the building's main elements. Whilst the floor area technique is presented here in the context of formal cost plan 1, some surveyors will also apply the technique when developing the OCE (as illustrated in Table 6.9), in an attempt to produce a more refined budget cost at the commencement of the project or in the case of refurbishment projects where finding comparable historic cost data becomes impossible.

### Example Cost Plan Based on Floor Areas (£/m² GIFA)

In this example, we will further develop the new school extension used earlier in the chapter during the development of the business case or the OCE for the project. To develop cost plan 1 (a parametric estimate) for the school extension project using floor areas, The BCIS database has been searched to find cost analyses that match as closely as possible the attributes of the proposed project, including:

- The overall size of the proposed building (900 m²)
- Nature of the work (horizontal extension)
- The number of storeys (two)
- The footprint of the building (area of each floor – 450 m²)
- The individual storey height (3 m)
- The main construction specification (steel frame, concrete floors, brick cladding)
- The anticipated level of quality (primary school, state sector)

Using these parameters in the BCIS database revealed 26 suitable projects. Further consideration of these would allow the quantity surveyor to select the most appropriate project, in this case a two-storey classroom block horizontal extension constructed in Yorkshire has been selected from which the partial cost plan shown in Table 6.14 has been developed. From the analysis you can see that the additional design information has allowed the quantity surveyor to establish a more detailed breakdown for the main construction works, breaking down the overall cost of the extension outlined in the OCE into elemental cost limits. This is achieved using the percentage distribution column located at the end of the historical cost data. Care is needed, though, to ensure the percentages are adjusted. In the example provided, the historic cost data for the building work totalled 52%, but we know the allowance for building work in our OCE is 100%, so some adjustments to the percentage distribution have been made.

Selection of the Yorkshire school teaching block extension was based on the many similarities in core areas such as specification, building function and layout and design quality. Looking at the historic cost data, it is constructed for a local authority and has a similar layout and construction specification to the client's proposed project. Additional areas of similarity include number of storeys and the comparable, although slightly larger, gross internal floor area. As such, although the Yorkshire project is slightly larger, located in Yorkshire and tendered in Q2, 2013, the project nevertheless provided the most appropriate comparator for the scheme under review.

**Table 6.14** Formal cost plan 1 for the school extension based on floor area (£/m$^2$).

| | Element | Total Cost | Cost per m$^2$ | EUQ | EUR | % |
|---|---|---|---|---|---|---|
| 0 | **Facilitating works** | **£0** | **£0** | | | **0** |
| 1 | **Substructure** | **£173,867** | **£193** | | | **12%** |
| 2 | **Superstructure** | **£608,535** | **£676** | | | **42%** |
| 2.1 | Frame | £86,933 | £97 | | | 6% |
| 2.2 | Upper floors | £57,955 | £64 | | | 4% |
| 2.3 | Roof | £72,445 | £80 | | | 5% |
| 2.4 | Stairs and ramp | £28,977 | £32 | | | 2% |
| 2.5 | External walls | £173,867 | £193 | | | 12% |
| 2.6 | Windows and external doors | £72,445 | £80 | | | 5% |
| 2.7 | Internal walls and partitions | £57,955 | £64 | | | 4% |
| 2.8 | Internal doors | £57,955 | £64 | | | 4% |
| 3 | **Internal finishes** | **£86,933** | **£97** | | | **6%** |
| 3.1 | Wall finishes | £28,977 | £32 | | | 2% |
| 3.2 | Floor finishes | £28,977 | £32 | | | 2% |
| 3.3 | Ceiling finishes | £28,977 | £32 | | | 2% |
| 4 | **Fittings, fixtures and equipment** | **£28,977** | **£32** | | | **2%** |
| 4.1 | Fittings, fixtures and equipment | £28,977 | £32 | | | 2% |
| 5 | **Services** | **£289,780** | **£612** | | | **38%** |
| 5.1 | Sanitary installations | £28,977 | £32 | | | 2% |
| 5.2 | Services equipment | £57,955 | £64 | | | 4% |
| 5.3 | Disposal installations | | £0 | | | |
| 5.4 | Water installations | £57,955 | £64 | | | 4% |
| 5.5 | Heat source | £72,444 | £80 | | | 5% |
| 5.6 | Space heating and air conditioning | £72,444 | £80 | | | 5% |
| 5.7 | Ventilation | £57,955 | £64 | | | 4% |
| 5.8 | Electrical installations | £86,933 | £97 | | | 6% |
| 5.9 | Fuel installations | | £0 | | | |
| 5.10 | Lift and conveyor installations | £28,977 | £32 | | | 2% |
| 5.11 | Fire and lightning protection | £28,977 | £32 | | | 2% |

*(Continued)*

**Table 6.14** (*Continued*)

|  | Element | Total Cost | Cost per m$^2$ | EUQ | EUR | % |
|---|---|---|---|---|---|---|
| 5.12 | Communications and security installations | £28,977 | £32 |  |  | 2% |
| 5.13 | Specialist installations | £28,977 | £32 |  |  | 2% |
| 5.14 | Builder's work in connection with services |  | £0 |  |  |  |
|  | **Subtotal** | **£1,448,892** | £0 |  |  | **100%** |

EUQ – elemental unit quanity; EUR – elemental unit rate.

**Table 6.15** Formal Cost Plan 1 for the school extension based on floor area (£/m$^2$).

|  | Element | Total Cost | Comments |
|---|---|---|---|
| **0** | **Facilitating works** | **£0** | Based on new extension – 900 m$^2$ |
| **1** | **Substructure** | **£173,867** | Based on new extension – 900 m$^2$ |
| **2** | **Superstructure** | **£608,535** | Based on new extension – 900 m$^2$ |
| 2.1 | Frame | £86,933 |  |
| 2.2 | Upper floors | £57,955 |  |
| 2.3 | Roof | £72,445 |  |
| 2.4 | Stairs and ramp | £28,977 |  |
| 2.5 | External walls | £173,867 |  |
| 2.6 | Windows and external doors | £72,445 |  |
| 2.7 | Internal walls and partitions | £57,955 |  |
| 2.8 | Internal doors | £57,955 |  |
| **3** | **Internal finishes** | **£86,933** | Based on new extension – 900 m$^2$ |
| 3.1 | Wall finishes | £28,977 |  |
| 3.2 | Floor finishes | £28,977 |  |
| 3.3 | Ceiling finishes | £28,977 |  |
| **4** | **Fittings, fixtures and equipment** | **£28,977** | Based on new extension – 900 m$^2$ |
| 4.1 | Fittings, fixtures and equipment | £28,977 |  |
| **5** | **Services** | **£289,780** | Based on new extension – 900 m$^2$ |
| 5.1 | Sanitary installations | £28,977 |  |

(*Continued*)

**Table 6.15** (*Continued*)

|  | Element | Total Cost | Comments |
|---|---|---|---|
| 5.2 | Services equipment | £57,955 | |
| 5.3 | Disposal installations | | |
| 5.4 | Water installations | £57,955 | |
| 5.5 | Heat source | £72,444 | |
| 5.6 | Space heating and air conditioning | £72,444 | |
| 5.7 | Ventilation | £57,955 | |
| 5.8 | Electrical installations | £86,933 | |
| 5.9 | Fuel installations | | |
| 5.1 | Lift and conveyor installations | £28,977 | |
| 5.11 | Fire and lightning protection | £28,977 | |
| 5.12 | Communications and security installations | £28,977 | |
| 5.13 | Specialist installations | £28,977 | |
| 5.14 | Builder's work in connection with services | | |
| **6** | **Complete buildings and building units** | **0** | |
| **7** | **Work to existing buildings** | **397,396** | Based on existing building floor area of 396 m$^2$ |
| | **Subtotal** | **£1,585,488** | |
| **8** | **External Works** | **£365,279** | |
| 8.1 | Site preparation works | | |
| 8.2 | Roads, paths and pavings | £275,279 | Based on area of 1,817 m$^2$ |
| 8.3 | Soft landscaping, planting and irrigation systems | | |
| 8.4 | Fencing, railings and walls | | |
| 8.5 | External fixtures | | |
| 8.6 | External drainage | | |
| 8.7 | External services | £85,000 | Upgrades to existing provisions |
| 8.8 | Minor building works and ancillary buildings | £5,000 | New sectional garage |
| | **Building works estimate** | **£1,950,767** | |
| **9** | Main contractor's preliminaries | £195,077 | Allowed at 10% |
| **10** | Main contractor's overheads and profit | £0 | Included |
| 10.1 | Main contractor's overheads | £0 | Included |
| 10.2 | Main contractor's profit | £0 | Included |

(*Continued*)

**Table 6.15** (*Continued*)

| Element | Total Cost | Comments |
| --- | --- | --- |
| **Works cost estimate** | **£3,070,331** | |
| Design fees estimate | £300,418 | 14% allowed – 12% for designers and 2% for others based on D&B |
| **Risk allowances estimate** | **£407,710** | |
| Design development | £214,584 | Based on development from RIBA stage 2–4 |
| Construction risk | £85,834 | |
| Employer change | £53,646 | |
| Employer other | £53,646 | |
| **Cost Limit (excluding Inflation)** | **£3,261,682** | |
| Tender inflation estimate | £277,569 | Assumed tendering in 3Q 2023 so (357–329/329*100) = 8.51% |
| **Cost Limit (excluding Construction Inflation)** | **£3,539,252** | |
| Construction inflation estimate | - | Fixed price contract so no allowance included. |
| **Cost Limit including Inflation** | **£3,539,252** | |
| VAT assessment | £707,850 | Client requested the addition of VAT @ 20% |

As noted above this is only a partial cost plan because this project consisted of a new build extension alongside the refurbishment of the existing school. Earlier in the chapter when the business case was presented, this was provided in two clear sections. Section A outlined the costs of the extension, with building work budgeted at £1,448,892 (see Table 6.7) and the refurbishment and external works costed at an additional £1,004,625 (see Table 6.8). Table 6.14 breaks down the OCE for the extension into elemental units; however, as the floor area for the existing building is different to the floor area for the extension, the cost per square metre column and percentage column have been removed. The refurbishment element of the scheme, along with other costs have not been included.

Additionally, the issue of market change has been addressed in Table 6.15 through the addition of inflation allowances for tender price movements from the current time until the anticipated tender and construction periods. Prior to inclusion in the earlier tables, the data had already been subject to rebasing. Two final remarks regarding Table 6.16: firstly, there is no allowance for construction inflation, as it is assumed this will be a fixed price contract (see chapter 5 for further details of this); and secondly, the tricky issue of VAT has been addressed. NRM1 in measurement rule 3.20.1 and 3.20.2 (RICS 2013, p. 65) advises that VAT should be left off the cost plan (cost plan produced

**Table 6.16** Formal Cost Plan 2 for a hotel using elemental unit rates and quantities.

| | Element | Total Cost | Cost per m$^2$ | EUQ | Unit | EUR |
|---|---|---|---|---|---|---|
| 0 | **Facilitating works** | **£0** | **£0** | | | |
| 1 | **Substructure** | **£803,600** | **£96** | | | |
| | Excavation, ground beams, filling to levels, lift pits, ground slab | £529,200 | £63 | 445 | m$^2$ | £1,189 |
| | Rotary bored piles | £215,600 | £26 | 445 | m$^2$ | £484 |
| | Under slab drainage | £58,800 | £7 | 445 | m$^2$ | £132 |
| 2 | **Superstructure** | **£4,586,400** | **£546** | | | |
| 2.1 | Frame | £313,600 | £37 | 445 | m$^2$ | £705 |
| 2.2 | Upper floors | £1,724,800 | £205 | 1780 | m$^2$ | £969 |
| 2.3 | Roof | £254,800 | £30 | 445 | m$^2$ | £573 |
| 2.3.1 | Roof structure | | | | | |
| | Flat roof, pre-cast concrete roof slab | £313,600 | £37 | 445 | m$^2$ | £705 |
| | Extra over for forming upstands and copings | £19,600 | £2 | 1 | Item | £19,600 |
| 2.3.2 | Roof covering | | | | | |
| | Single ply roof membrane, insulation, rainwater outlets | £196,000 | £23 | 445 | m$^2$ | £440 |
| | Mansafe system | £9,800 | £1 | | Item | |
| 2.4 | Stairs and ramp | £78,400 | £9 | 2 | nr | £39,200 |
| 2.5 | External walls | £940,800 | £112 | 1578 | m$^2$ | £596 |
| 2.6 | Windows and external doors | £284,200 | £34 | 250 | m$^2$ | £1,137 |
| 2.7 | Internal walls and partitions | £205,800 | £25 | 1160 | m$^2$ | £177 |
| 2.8 | Internal doors | £245,000 | £29 | 90 | nr | £2,722 |
| 3 | **Internal finishes** | **£964,124** | **£115** | | | |
| 3.1 | Wall finishes | £532,924 | £63 | 5500 | m$^2$ | £97 |
| 3.2 | Floor finishes | £235,200 | £28 | 1980 | m$^2$ | £119 |
| 3.3 | Ceiling finishes | £196,000 | £23 | 1980 | m$^2$ | £99 |
| 4 | **Fittings, fixtures and equipment** | **£1,155,714** | **£138** | | | |
| 4.1 | Fittings, fixtures and equipment | £1,155,714 | £138 | 64 | nr | £18,058 |
| 5 | **Services** | **£4,290,342** | **£511** | | | |
| 5.1 | Sanitary installations | £340,942 | £41 | 1800 | m$^2$ | £189 |
| 5.2 | Services equipment | £1,323,000 | £158 | | | |
| 5.3 | Disposal installations | £49,000 | £6 | 64 | nr | £766 |
| 5.4 | Water installations | £176,400 | £21 | 1800 | m$^2$ | £98 |
| 5.5 | Heat source | £793,800 | £95 | 1800 | m$^2$ | £441 |
| 5.6 | Space heating and air conditioning | | | | | |
| 5.7 | Ventilation | | | | | |

*(Continued)*

**Table 6.16** (*Continued*)

| | Element | Total Cost | Cost per m$^2$ | EUQ | Unit | EUR |
|---|---|---|---|---|---|---|
| 5.8 | Electrical installations | £1,029,000 | £123 | 1800 | m$^2$ | £572 |
| 5.9 | Fuel installations | | | | | |
| 5.1 | Lift and conveyor installations | £303,800 | £36 | 2 | nr | £151,900 |
| 5.11 | Fire and lightning protection | £19,600 | £2 | 1800 | m$^2$ | £11 |
| 5.12 | Communications and security installations | £235,200 | £28 | 1800 | m$^2$ | £131 |
| 5.13 | Specialist installations | | | | | |
| 5.14 | Builders work in connection with services | £19,600 | £2 | 1800 | m$^2$ | £11 |
| | **Subtotal** | **£11,800,180** | **£1,405** | | | |
| 8 | **External works** | **£1,195,600** | **£142** | | | |
| 8.1 | Site preparation works | £539,000 | £64 | | | |
| 8.2 | Roads, paths and pavings | | | | | |
| 8.3 | Soft landscaping, planting and irrigation systems | | | | | |
| 8.4 | Fencing, railings and walls | | | | | |
| 8.5 | External fixtures | | | | | |
| 8.6 | External drainage | £382,200 | £46 | | | |
| 8.7 | External services | £274,400 | £33 | | | |
| 8.8 | Minor building works and ancillary buildings | £0 | £0 | | | |
| | **Building works estimate** | **£12,995,780** | **£1,547** | | | |
| 9 | Main contractor's preliminaries | £1,689,451 | £201 | 13% | % | |
| 10 | Main contractor's overheads and profit | | | | | |
| 10.1 | Main contractor's overheads | | | | | |
| 10.2 | Main contractor's profit | | | | | |
| | **Works cost estimate** | **£14,685,231** | **£1,748** | | | |
| | Design fees estimate | £1,762,228 | £210 | 12% | | |
| | **Risk allowances estimate** | **£2,055,932** | **£245** | | | |
| | Design development | £734,262 | £87 | 5% | | |
| | Construction risk | £440,557 | £52 | 3% | | |
| | Employer change | £440,557 | £52 | 3% | | |
| | Employer other | £440,557 | £52 | 3% | | |
| | **Cost limit (excluding inflation)** | **£18,503,392** | **£2,203** | | | |
| | Tender inflation estimate (4Q22) | £1,069,496 | £127 | 5.78% | | |
| | **Cost limit (excluding construction inflation)** | **£19,572,888** | **£2,330** | | | |
| | Construction inflation estimate (2Q23) | £336,654 | £40 | 1.72% | | |
| | **Cost limit including inflation** | **£19,909,541** | **£2,370** | | | |
| | **VAT assessment** | Excl | | | | |

excluding VAT), and furthermore that specialist advice is sought on by the client in relation to their individual exposure to VAT given the interplay of capital allowances, exceptions and VAT within the tax system. For business organisations. VAT has been added to this example to allow this discussion to be included and to demonstrate the financial impact of taxation.

Once these various elements are added into the cost plan in Table 6.15, we can see the overall cost of the scheme is £3,539.252 (excluding VAT) or £4,247,102 Including VAT.

### 6.4.3   Formal Cost Plan 2

The preparation of a formal cost plan is progressive and highly dependent on the amount of design information available to the quantity surveyor at the time. As a result, it is difficult to identify which cost planning techniques the quantity surveyor will be able to use and when; this will be largely governed by the speed of design and the method of procurement the employer decides to adopt. NRM1 suggests Formal Cost Plan 2 is produced at stage 3 *Spatial Coordination* in the RIBA plan of work. At this stage, the RIBA (2020) suggests design development detailed design studies are undertaken to make sure the building is spatially coordinated; this stage ends with the submission of a planning application for the development. From a costing perspective, the RIBA plan of work suggests that cost exercises are undertaken on the design and specification with the financial implications of achieving project outcomes established, whilst the cost plan is 'iteratively updated with increasing levels of cost certainty as greater detail of the architectural proposal is developed' (RIBA 2020, p. 51).

The mapping of design development presented in Table 6.11 suggests the design will have been developed to somewhere between 30% and 50% at this stage. As such, the RICS practice note (RICS 2020) advises the adoption of *semi-detailed unit costs*. Within the NRM framework, it can be reasonably expected that the surveyor will look and apply these semi-detailed unit costs using elemental unit quantities (EUQs) supported by elemental unit rates (EURs). In this context, the measure and pricing of the project is likely to take place at either sub-element or component level depending on the extent of design development for that aspect of the project.

The first stage in this process would be for the surveyor to develop EUQs at element, sub-element or component level, depending on levels of design development. These should be measured using the *tabulated rules of measurement* provided in part 4 of the NRM (RICS 2013, pp. 69–327). EURs can range from calculating the area serviced by a heating system based on net floor area down to counting the number of sanitary fittings in the building. The surveyor would ultimately seek to measure each major type of the element under consideration. In terms of floor finishes, for example, ceramic/porcelain tiles, carpets, non-slip sheet flooring and laminate or real wood flooring would all be measured separately.

The next stage in the process is to then develop unit rates for these elements, in the form of an elemental unit rate. Also known as composite rates, these rates can be sourced from a range of historic project data, commercial price books or through soft market testing depending on the complexity of the project under review. Composite rates are really a collection of rates that have been brought together to provide the surveyor with a *semi-detailed unit cost* that is more accurate than the floor area rates adopted earlier, whilst also reflecting the ongoing lack of complete information that would make accurate measurement and unit rate costing possible.

### Example Cost Plan Based on Elemental Rates

The cost plan illustrated in Table 6.16 relates to a hotel development. The EUQs and EURs provided in the cost plan have been developed based on a hotel cost model obtained from a commercially available price book. The cost model is based on an 8,400 m$^2$ hotel in Manchester. Finally, the rates in the example have been updated for price movements and tender and construction inflation added based on the project being tendered via design and build in 4th Quarter 2022 with the construction mid-point occurring in 2nd Quarter 2023.

From the cost plan illustrated in Table 6.16 it is notable that additional design information has allowed the surveyor to establish a more detailed breakdown for the main construction works with some partial quantities established from the information produced by the design team. In some parts, for instance substructure and roof covering, the example has considered component level cost components given the level of design information provided for these areas of the project.

Adding this data, along with the information to the summary elemental cost plan, it is now possible to prwovide the client with a more detailed analysis of the project's financial performance and constraints as illustrated in Table 6.17. In practice, this summary would be supported by extensive analysis and supporting build-up calculations.

**Table 6.17** Approximate quantities for internal walls and partitions.

| | Element | Total Cost | Cost per m$^2$ | Approx. Quant | Unit | Composite Rate |
|---|---|---|---|---|---|---|
| **2** | **Superstructure** | **£4,586,400** | **£546** | | | |
| 2.7 | Internal walls and partitions | £205,800 | | | | |
| 2.7.1 | Walls and partitions | | | | | |
| 2.7.1.1 | Internal walls; blockwork, 100 thick | £93,728 | | 1600 | m$^2$ | £58.58 |
| 2.7.1.1 | Internal walls; blockwork, 140 thick | £1,111 | | 25 | m$^2$ | £44.45 |
| 2.7.1.2 | Extra over internal walls for forming openings in walls for doors and the like door opening 1000 mm | £210 | | 12 | Nr | £18.00 |
| 2.7.1.3 | Fixed partitions, acoustic metal stud, 100 thick | £29,581 | | 714 | m$^2$ | £41.43 |
| 2.7.1.4 | Extra over fixed partitions for forming openings in walls for doors and the like door opening 1000 mm | £2,044 | | 146 | Nr | £14.00 |

### 6.4.4 Formal Cost Plan 3

As mentioned earlier, the preparation of a formal cost plan is an iterative process and one that is highly dependent on information for accuracy and detail. Assuming the project is procured using traditional procurement, the final reporting stages outlined in NRM1 is Formal Cost Plan 3, whilst other reports are likely to be produced, such as pre-tender estimates, contract sum analysis based on submitted tenders and others. Formal cost plan 3 is the final formal cost plan reporting stage NRM1 identifies. This stage of reporting takes place at the end of stage 4 *Technical Design* in the RIBA plan of work.

The RIBA Plan of Work (RIBA 2020, p. 58) suggests at this stage of design development, the focus of the design team will have shifted towards providing the detailed technical information including specifications and drawings required to manufacture and construct the building with the end of the stage marked by the tendering process. Once again, from a costing perspective, the RIBA plan of work suggests detailed elemental analysis of cost, together with a full bill of quantities, unit cost items or pricing schedules to be produced. The final cost plan produced at this stage will also go on to become a pre-tender cost estimate for the client (RIBA 2020, p. 58).

Given that design maturity will have, through this final design phase, achieved maturity level 4 and as tender documents are prepared at maturity level 5, Formal Cost Plan 3 will usually be produced using approximate quantities to provide the client with the most accurate pre-tender cost forecast for the project. This is unlike earlier cost plans, which may have been derived from the overall budget, or known costs for similar completed projects and are refined by the surveyor using their own experience and professional judgement. An elemental cost plan produced using approximate quantities will represent the first major complete attempt to measure defined quantities from the drawing (or to extract them from a building information model) using the *tabulated rules of measurement* contained in part 4 of the NRM. Given the levels of quantification achievable at this stage in the project, this elemental cost plan will provide the design team and client with a far more accurate picture of both the overall cost of the project and the distribution of costs through the various elements, sub-elements and components that make up the building. In reality, this document will almost constitute a full bill of quantities for the project, with detailed takeoffs fully supporting the dimensions reported. It is normal for this document to contain hundreds of pages of measurement and cost data at this stage in the project's development.

**Example Cost Plan Based on Approximate Quantities**

Referring once again to the 8,400 m$^2$ hotel in Manchester, Table 6.17 provides an excerpt from a comprehensive cost plan for the project developed, using approximate quantities. Given the complexity and scale of a cost plan prepared using approximate quantities it is impossible to provide a fully detailed example. Consequently, excerpts from the larger cost plan have been provided in Tables 6.17 and 6.18 to illustrate the process and the levels of detail required. Once again, the quantities provided are derived from a hotel cost model published in a commercially available price book with the rates updated to take account of tender price inflation since the publication of the text.

**Table 6.18** Approximate quantities for sanitary installations.

| | Element | Total Cost | Cost per m$^2$ | Approx. Quant | Unit | Composite Rate |
|---|---|---|---|---|---|---|
| 5 | **Services** | **£976,160** | **£439** | | | |
| 5.1 | Sanitary installations | £279,230 | £126 | | | |
| 5.1.1 | Sanitary appliances | | | | | |
| 5.1.1.1 | WC | £1,200 | | 8 | Nr | £150 |
| 5.1.1.1 | Urinals | £988 | | 4 | Nr | £247 |
| 5.1.1.1 | Wash basin | £1,020 | | 6 | Nr | £170 |
| 5.1.1.1 | Belfast sink | £822 | | 3 | Nr | £274 |
| 5.1.2 | Pods | | | | | |
| 5.1.2.3 | Shower room pod – details to follow from hotel franchise | £275,200 | | 64 | Nr | £4300 |

## 6.5 Summary

In summary, this chapter of the textbook has shown how pre-contract financial management develops through the option selection and design phases of the project. The chapter has shown the phases of pre-contract financial management implemented by the cost consultant/quantity surveyor from the very outset of the project, commencing with the 'order of cost estimate' and an option appraisal, both of which form a key part of the employer's overall appraisal of the project. The OCE is the 'determination of possible costs of a building early in the design stage in relation to the employer's functional requirements' (RICS 2013, p. 14). The OCE is prepared at Stage 1: Preparation and brief of the *RIBA Plan of Work 2020*. Whilst allowing the client to evaluate potential options, locations or levels of refurbishments for a given building, the OCE also establishes the project cost limit for the employer.

As the design develops, the cost consultant/quantity surveyor will move forwards with their financial management of the pre-contract stages of the project. At this stage cost plans will be produced. As evidenced earlier, a cost plan provides a statement of how the available budget will be allocated to the various elements of the building. It provides a frame of reference to be used when developing the design to ensure that costs are fully controlled as the project moves forwards through the RIBA work stages. It has also been shown that cost planning is performed in stages of increasing detail as more design information becomes available (RICS 2013, p. 50). With formal cost plans prepared in the following stages of the *RIBA Plan of Work 2020*:

- Stage 2: Concept Design      Formal Cost Plan 1
- Stage 3: Spatial Coordination      Formal Cost Plan 2
- Stage 4: Technical Design      Formal Cost Plan 3

Throughout this process the surveyor will forecast and predict project costs; however, this process is reliant on design information provided by the design team and the levels of design development attained. The main purposes or benefits of cost planning are revealed in the *New Rules of Measurement* (RICS 2013, p. 50) to include:

- Ensuring that employers are provided with value for money;
- Making employers and designers aware of the cost consequences of their desires and/or proposals;
- Providing advice to designers that enables them to arrive at practical and balanced designs within budget;
- Keeping expenditure within the cost limit approved by the employer; and
- Providing robust cost information upon which the client can make informed decisions.

## References

Association of Project Management (2021). What is a business case? https://www.apm.org.uk/resources/what-is-project-management/what-is-a-business-case/#:~:text=Definition, rationale%20for%20the%20preferred%20solution, (accessed 29 May 2021)

Babangida, I., Olubodun, F., and Kangwa, J. (2012). Building refurbishment: Holistic evaluation of barriers and opportunities. In: Procs 28th Annual ARCOM Conference, 3–5 September 2012 (ed. S.D. Smith), 1289–1298. Edinburgh, UK: Association of Researchers in Construction Management,

BCIS (2021). Economic significance of maintenance 2021. https://service-bcis-co-uk.salford.idm.oclc.org/BCISOnline/Briefings/EconomicBackground/3347?returnUrl=%2FBCISOnline%2FBriefing&returnText=Go%20back%20to%20briefing%20summary&sourcePage=Help (accessed 29 May 2021).

Constructing Excellence (2006). Rethinking standards in Construction. https://constructingexcellence.org.uk/wp-content/uploads/2015/02/rethinkingstandardsbrochurefinal.pdf (accessed 29 May 2021).

Egbu, C. (1996). Characteristics and difficulties associated with refurbishment, Construction papers No. 66. Ascot: Chartered Institute of Building.

Higham, A., Bridge, C., and Farrell, P. (2017). *Project Finance for Construction*. Abingdon: Routledge.

HM Treasury (2018). Guide to developing the project business case. https://assets.publishing.service.gov.uk/government/uploads/system/uploads/attachment_data/file/749086/Project_Business_Case_2018.pdf (accessed 29 May 2021).

RIBA. (2020). RIBA Plan of work 2020. https://www.architecture.com/-/media/GatherContent/Test-resources-page/Additional-Documents/2020RIBAPlanofWorktemplatepdf.pdf (accessed 29 May 2021).

RICS. (2013). *NRM1: Order of Cost Estimating and Cost Planning for Capital Building Works*, 2nd ed. Coventry: RICS.

RICS. (2018). Code of measuring practice, 6th ed. https://www.rics.org/globalassets/rics-website/media/upholding-professional-standards/sector-standards/valuation/code-of-measuring-practice-6th-edition-rics.pdf (accessed 29 May 2021).

RICS (2020). Professional statement: Cost prediction. https://www.rics.org/globalassets/rics-website/media/upholding-professional-standards/sector-standards/construction/cost-prediction_ps_1st-edition-19-nov-2020.pdf (accessed 29 May 2021).

Wolstenholme, A. (2009). Never waste a good crisis. https://constructingexcellence.org.uk/wp-content/uploads/2014/10/Wolstenholme_Report_Oct_2009.pdf (accessed 29 May 2021).

# 7

# Financial Management

Life Cycle Costing

## 7.1 Introduction

Chapter 7 looks at the growing importance of life cycle cost analysis in view of the longer-term interest that clients and tier one contractors are now taking in buildings, as contract award criteria continue on the paradigm shift in focus away from a lowest-price-wins mentality to one led by a longer-term focus on value and, importantly, the trade-off between the initial capital costs and longer-term operational costs of the asset or component, especially in the public sector. As this balance continues to play out in the marketplace, the importance of how future maintenance costs are forecast is also changing. However, further effort is required if the deeply ingrained business culture that compartmentalises capital and maintenance funding is to be overcome on traditional projects.

## Part 1 – The Case for Life Cycle Analysis

Traditionally, little thought would be given to the operational needs of the asset post-completion, with capital and revenue budgets often sitting with different budget holders with little evidence of trade-offs between capital spend and revenue spend occurring. As a result, those charged with delivering new assets for the organisation would focus on getting the most from capital budget in terms of numbers of projects, rather than ensuring the quality of the assets developed. This focus fitted well with the dominant use of a lowest-price-wins procurement mentality. Regrettably, this aggressive focus on cost minimisation often led to low quality assets. Contractors bidding at cost or even below cost to secure work adopted an aggressive commercial model focused on extracting their margin through, inter alia, aggressive value engineering focused on cost reduction (Hackett 2020). Such an aggressive focus on cost at the expense of quality consequently impacted on the client's revenue budget, as it would always led to higher maintenance cost exposure over the asset's life cycle (Swaffield and McDonald 2008).

More recently, over the last 15 years, the sector has seen a pragmatic shift in both clients' mindsets and an associated change in focus at most major, tier one contractors to one where quality is seen as paramount and the delivery of long-term value is seen as

*Introduction to Built Asset Management*, First Edition. Dr Anthony Higham, Dr Jason Challender, and Dr Greg Watts.
© 2022 John Wiley & Sons Ltd. Published 2022 by John Wiley & Sons Ltd.

more important than simply delivering cheap projects. As part of this shift in perspective, it is becoming increasingly clear to project stakeholders that the careful balancing of capital and operational costs can deliver long-term value for money benefits for the client.

This is especially important from a maintenance perspective, where control and predication of cost profiles during the design and construction phase of the building can play out into stronger financial management of the asset during its operational life. Lessons learnt from the UK's dominant position in the private financing initiative (PFI) market have demonstrated that proactive life cycle cost modelling is an essential feature of effective investment decision-making (Miller 2014) to ensure the income flows from the PFI asset cover likely maintenance and upkeep costs, with most PFI contractors seeking to expend a maximum of 2% of capital cost per annum on maintenance over the 30–40-year concession. Given this increase in focus towards operational (life cycle) expenditure, this chapter looks to introduce the principles and processes associated with life cycle costing.

## 7.2   Forecasting Financial Impacts of Building Maintenance

Writing this book from a background as a quantity surveyor, it is strongly apparent that the majority of texts looking at the financial management of buildings place emphasis on the management and prediction of capital costs and little focus on anything that happens post-construction. Yet this perspective is evidently reinforced in practice, with the Royal Institution of Chartered Surveyors (RICS) professional services contract for the appointment of financial experts (quantity surveyors) obligating the management of the capital cost of the project to be an essential service, whereas life cycle cost modelling is merely an optional addition. Despite this, it is also acknowledged in literature that the operational costs of a building over its lifetime are significantly higher than the capital construction costs. Constructing Excellence (2004) has suggested the operational costs of a typical office building over 30 years will be five times higher than its capital construction cost. The exact magnitude is open to some debate. Ive (2006), for instance, has suggested a magnitude of 1:1.5 would be accurate, a position supported by BCIS (2011), who analysed a 763 m$^2$ school dining hall over its anticipated 20-year lifespan. The analysis shown in Table 7.1 illustrates that when the costs of constructing the school dining room are considered against occupancy costs, the dining room would cost 1.98 times more to 'occupy'. Although some warn against the use of such ratios, suggesting that in reality it is really impossible to compute a clear ratio of cost, as there are a myriad of other factors that are likely to significantly distort any analysis. There remains a universal acceptance that the maintenance and occupancy costs of a building will be significantly higher than the costs of initial construction. On this basis, it seems appropriate that decision makers should proactively consider maintenance costs and appraise the value benefits of spending slightly more on the asset's initial construction to reduce through life costs.

The benefits of proactively considering the through life costs of the asset at the early stages of design development have been demonstrated in the PFI sector where investors and construction contractors came face to face with the need to consider the balance between capital and operational costs from the perspective of value. The operation of this decision-making process has been mapped out in detail by Miller (2014) in an article

**Table 7.1**  Occupancy Costs for a 763m$^2$ school dining room.

| Element | Initial cost (£) | Future occupancy costs (£) |
|---|---|---|
| **Total Initial Capital Costs** | 881,865 | |
| **Annual Costs** | | |
| 2.1 Major Replacement | | 928 |
| 2.3 Redecorations | | 294 |
| 2.4 Minor Repairs, etc. | | 1,723 |
| 2.5 Unscheduled repairs, etc. | | 173 |
| 2.6 Grounds maintenance and external works | | 137 |
| 3.1 Cleaning | | 3,023 |
| 3.2 Utilities | | 1,931 |
| 3.3 Administrative costs | | 2,695 |
| 3.4 Overheads | | 547 |
| **Total Average Annual Costs (£/100m$^2$)** | | **11,452** |
| **Projected Total (over 20 years)** | | **1,747,647** |
| **Ratio (Running costs: Capital Costs)** | | **1.98** |

*Source:* Adapted from British Standards Institution (2011). BS8534:2011 Construction procurement policies, strategies and procedures – Code of practice. London: British Standards Institute.

published in *Maintenance and Engineering* where he explained that life cycle modelling was a key feature of investment decision-making and would be subject to careful scrutiny by not only the SPV (special purpose vehicle) but also by end-user clients and funders. Although these stakeholders would approach the model with different objectives, for instance, the SPV would be focused on the commercial aspects of the data in terms of balancing income against outgoings, whereas funders would be looking at the risk the project presented from the perspective of whether the life cycle commitments balance attractively with the income the investment would generate. As such, the life cycle analysis is often reduced to a simple expression of total life cycle cost/gross floor area/years of concession (£/m$^2$/year). Regardless of the perspective adopted, the PFI market demonstrates the clear benefits to the decision maker. Whilst in this context it has to be acknowledged that life cycle costing is used to anticipate maintenance and renewal costs and thereby to correctly predict annual charges to the end-user (Miller 2014), the benefits associated with predicting maintenance and renewal costs to those involved in asset management must be acknowledged.

Despite the benefits observed in the PFI market, there continues to be a lack of commitment to the use of life cycle costing, both in public and private sectors (Higham et al. 2015). It is, however, noteworthy that government policy has continued to shift away from the adoption of a price-led, CAPEX-oriented perspective on projects. Even though capital cost remains an important yardstick of success, how financial performance is reviewed continues to evolve in government policy. This shift in focus can be traced back over the last 20 years. *The Green book*: *Appraisal and Evaluation in Central Government* published in 2003 presents the first policy framework to advocate the adoption of a value for money perspective. This initial guidance asserted that value was to be measured through the ratio between functionality and whole-life cost of the built asset.

The importance of life cycle costing was again reinforced through the publication of the government's vision for the evolution of the construction industry captured in its strategic policy document *Construction 2025: Industrial Strategy for Construction – Government and Industry in Partnership*. The policy links to other government policies such as Digital Nations and the development of digital information management in construction through Level 2 and 3 BIM models which run for the asset's full life cycle, together with the government's commitment to an eased transition from construction to occupancy through a policy called 'soft landings', where there will be a three-year transition between the construction team and asset management team. At the core of the industrial strategy, however, is the government's commitment to 'a 33% reduction in the initial cost of construction and the whole-life cost of built assets' (HM Government 2013).

The latest position on life cycle analysis can be seen through the *Construction Sector Deal* (HM Government 2018) with both policies emphasising the desire to achieve, amongst other things, a 33% cost reduction by 2025. It must, however, be emphasised that both policy documents advocate that the 33% target is not to be simply achieved through capital cost savings, but through the adoption of a life cycle perspective. This is further emphasised in the Construction Sector Deal, which promotes the need for a 'shift in focus from the costs of construction to the costs of a building across its life cycle'. This, the policy suggests, will occur within a procurement strategy focused on whole-life asset value. These policies ultimately then link in with other government policies such as Digital Nation and the development of digital information management in construction through Level 2 and 3 BIM models which run for the asset's full life cycle, together with the government's commitment to an eased transition from construction to occupancy through a policy called 'soft landings' where there will be a three-year transition between the construction team and asset management team. From this policy framework it remains clear that the government seeks to advocate the need for a shift in viewpoint away from a short-sighted CAPEX and OPEX perspective of buildings associated with the traditional management of a capital and revenue budget, to an integrated cost model based on total expenditure (TOTEX) and value over the asset life cycle.

It is hard to argue that the importance of life cycle costing is not limited to public sector clients, although such clients account for 40% of total industry GDP. The majority of private sector clients have very different construction motivations, with decision marking typically led by a short-term perspective focused on accounting ratios such as return on capital employed thus ensuring shareholder satisfaction, a perspective that ultimately continues to support the importance of capital cost and price to project delivery. It is hoped the benefits of adopting a life cycle value perspective, whereby construction and occupancy perspectives are integrated into one continued perspective, will become increasingly evident to the wider market and adoption by private sector clients will ultimately follow.

## 7.3   Defining Life Cycle Costing

When you start to read around life cycle cost analysis, or attend lectures at university or industry-based CPD events, you will notice that a number of terms are used interchangeably to describe the concept under consideration. Arguably the most frequent terms you will have heard mentioned are LCC (life cycle costing) and WLC (whole-life costing); these will either be used through the presentation or book to refer to the same concept, or, to reinforce the confusion, the speaker or lecturer will adopt each term at different stages of the lecture/presentation but will be explaining the same concept or principle. However, in the realities of practice, LCC and WLC are very different concepts and, importantly, one adopts a much wider, investment-led perspective. Whilst this can be important to those working in the asset management sector, the apparent confusion and misuse of the terms could lead to the commissioning client receiving something inferior to the desired outcome. There are fundamental differences between LCC and WLC. With the former relating exclusively to financial management of the asset from a perspective of cost prediction and the latter (WLC) providing a much wider perspective of the asset with modelling of occupancy costs and revenue streams considered.

In an attempt to allay some of the confusion around the meaning of these different terms, this section of the chapter will define and explain each of the concepts adopting the international definitions provided by (ISO 15686-5:2017) and reaffirmed in the authoritative guidance provided by the RICS via the NRM (*New Rules of Measurement*) captured below.

---

*Life Cycle Cost* – The cost of an asset, or its parts throughout its life cycle, whilst fulfilling the performance requirements.

*Life Cycle Costing* – A methodology for the systematic economic evaluation of life cycle costs over a period of analysis, as defined in the agreed scope of assessment.

*Whole-Life Cost* – All significant and relevant initial and future costs and benefits of a building, facility or an asset, throughout its life cycle, whilst fulfilling the performance requirements.

*Whole-Life Costing* – A methodology for the systematic economic evaluation used to establish the total cost of ownership, or the whole-life costing of option appraisals. It is a structured approach addressing all costs in connection with a building or facility (including construction, maintenance, renewals, operation, occupancy, environmental and end of life). It can be used to produce expenditure profiles of a building or facility over its anticipated life span or defined period of analysis.

---

As can be seen from the definitions above, whole-life costing is a far more comprehensive level of analysis than that provided as part of a typical life cycle costing appraisal. As illustrated in Figure 7.1, the focus of the life cycle costing is limited to considering the expenditure profile of the building with a focus on replacement and maintenance of the elements, sub-elements and components that make up the building. Whereas whole-life costing

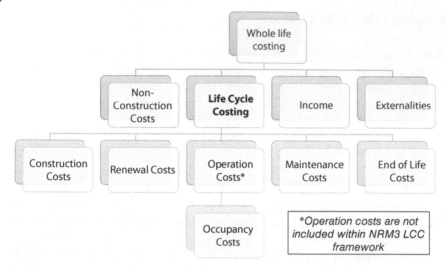

**Figure 7.1** Cost categories for LCC and WLC analysis. *Source*: RICS 2014, p. 23.

(WLC) considers not only the expenditure profile but also the potential income streams, the impact of externalities and other non-construction–related expenditure associated with the building. For most construction professionals, developing such a complex overview of an asset would sit outside their immediate competence and would thus require the input of a series of professionals with expertise in the various elements of the model. For instance, income predictions would require the expertise of a registered valuer, whereas costing and maintenance demands could be modelled by an asset manager working alongside building and quantity surveying colleagues.

Figure 7.1 identifies the four principal cost centres associated with developing a whole-life cost model for a building; these cost centres are each explained to give you a full appreciation of what would typically be included in a WLC analysis.

### 1. Non-Construction Costs

Non-construction costs cover any additional costs the building owner is likely to incur over the lifetime of the asset, or over the period the study is evaluating. Non-construction costs is a broad cost centre that incudes expenditure incurred by the client that falls outside the scope of construction costs managed by the surveyor. Examples of non-construction costs included within this cost centre could include:

- *Site Costs* – The costs associated with purchasing the land or the existing building(s) occupying the site. The transactional costs such as legal fees, taxation or auction costs are also likely to be included in this part of the cost analysis.
- *Finance Costs* – This cost centre includes any interest, bank charges or other fees the client pays in relation to funding the project.
- *Rental Costs* – If the client is seeking to lease out the building. The client's costs including legal and agent fees on the initial letting, void costs and other associated outlay would be included in this cost centre.

### 2. Life Cycle Cost

The life cycle cost of the project covers all the standard expenditure associated with a building, these can include:

- Construction costs;
- Renewal costs;
- Operation and occupancy costs;
- Maintenance costs; and
- End-of-life costs (demolition costs and alike).

The cost centres included in this aspect of the appraisal will be largely dependent on the client's requirements and the time horizon over which the study is to be conducted.

### 3. Income

Determining the project's viability is normally the preserve of the development/valuation/real estate surveyor. As part of the business case (typically a residual valuation) for the development they will provide detailed advice to clients about the overall development budget this would include some appraisal of the potential income generation likely to emerge from a project. This appraisal could include:

- Income from the sale of either the land or the building at the end of its design life;
- Income from a third party during the building's occupation, i.e. rental return.

It is important to also remember that income can be negative, to reflect lost income, which is likely to occur over the life cycle of the building, especially if it is rented as the asset will potentially have periods of vacancy (void periods) where the building owner could incur costs such as security for the asset along with exposure to potential costs such as business rates. As result, the whole-life cost analysis needs to account for periods when the building is vacant or undergoing refurbishment.

### 4. Externalities

The final cost centre is 'externalities'. These are costs associated with the asset that are not necessarily reflected in the transaction cost between the provider (constructor) and the consumer (client), for example, the social benefit arising from the building, or environmental damage created by the building and so on. Since the publication of the Social Value Act in 2012, public sector organisations have been required to measure and demonstrate the positive externalities their project generate, by way of a social value ratio. Over recent years this has become an important project success metric with a number of clients adopting social value as a key criterion for tender submissions but also for contract related key performance indicators (KPIs). Externalities such as social value have also now become a key marketing and PR tool for major construction organisations.

## 7.4 Challenges Associated with Life Cycle Prediction

Despite the many advantages associated with implementing life cycle cost or whole-life costing, the techniques are not without challenges. For some, the challenges begin and end with the mathematical nature of the technique and the challenges associated with applying such abstract concepts as the time value of money, a concept that will be explored later in the chapter. The real issue is, unfortunately, that accurate life cycle costing is heavily reliant on accurate and high-quality data, yet the challenges of both reliable data availability and data uncertainty are well recognised (Giuseppe et al. 2017). However, authors such as Oduyemi (2015) and Giuseppe et al. (2017) argue that these challenges are to be found in the mathematical base of LCC modelling, with the use of deterministic as opposed to probabilistic approaches to modelling responsible for these issues. Yet the gap between the academy and professional practice is vast. It is in academia, where issues associated with LCC are analysed and solutions discovered such as the one discussed here, rarely play out in practice. Instead, practice concerns itself with the operationalisation of LCC and, as such, the issues facing practitioners remain twofold. Firstly, data availability, and secondly, uncertainty.

The first issue is one of data and finding sufficient reliable data with which to undertake the analysis. Whilst some cost centres can be accurately predicted, such as capital costs of construction and to an extent the income/revenue the asset will generate, others, such as anticipated maintenance expenditure, are much more subjective. Some working in the sector producing life cycle analysis suggest a lot of the predictions cannot be made using reliable data and become more of an art in terms of forming educated predictions based on previous experiences; others in practice advocate that the use of historic maintenance data profiles from within the organisation, supported by manufacturers' product data can generate reasonably accurate predictions. Either way, it is evident for those producing life cycle cost models that data presents one of the major challenges impeding the technique. As yet, there is no readily available supply of commercial data in the way there is for capital cost prediction via publications such as those by BCIS or commercial price books such as Spon and Laxtons.

The second challenge facing those producing the life cycle model is the issue of uncertainty. Whilst uncertainty presents a challenge for anyone forming predictions, in construction these predictions are often short term, for instance with capital cost planning or the costing of maintenance based on condition surveys the time horizon is typically 2–3 years; whereas with a life cycle cost model the time horizon can be anything from 20–40 years. The levels of uncertainty associated with life cycle costing are depicted in Figure 7.2.

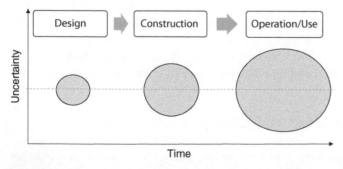

**Figure 7.2**   Uncertainty over time within life cycle cost models.

Figure 7.2 illustrates the levels of uncertainty at each of the key stages in the project life cycle. The figure attempts to depict the impact of time on the levels of uncertainty faced by the producer of the life cycle cost model. It is these levels of uncertainty that impact on the accuracy of the final output. Uncertainty is a key feature of most of the built environment. Unlike manufacturing, where products are mass produced after undergoing extensive prototyping and testing including simulated durability testing, the built environment consists of a series of unique structures, built without the benefit of prototyping or factory testing. As a result, uncertainty is embedded into the process from the very inception of the project at the initial business case stage and throughout the building's life cycle. Over time, this level of uncertainty grows. At the design stage uncertainties are often linked to issues such as specifications for elements of the building such as quality requirements for finishings. Other sources of uncertainty can include potential planning requirements or the overall design and flow of the building with financial impacts associated with floor-to-ceiling heights, for instance, feeding into the cost model.

As the project develops, additional levels of uncertainty develop as the construction phase commences. This uncertainty can come from issues such as ground conditions, energy costs, such as transportation costs, and economic variability in material and subcontractor pricing. For example, at the time of writing in mid-2021, the world was emerging from the COVID-19 pandemic with supply and demand mismatches as economies quickly recovered, leading to what the *Financial Times* called record-high commodity prices and the start of a supercycle where prices remain significantly above their long-term trend (Hume et al. 2021). The impact of this for construction has been significant rises in materials costs, which as one builder reported had put £6,000 on the cost of a loft conversion from November 2020 to April 2021.

Moving out to the longer term, and the occupancy phase of the building, the levels of uncertainty simply morph. At this point in the life cycle, uncertainties are introduced due to a variety of issues ranging from building users, social change, maintenance levels and economic data. Some of these issues will be explored later in the chapter. However, taking economic uncertainty, given the financial focus of the model, Giuseppe et al. (2017) observed that accuracy in life cycle modelling is heavily dependent on future trends in economic data such as changes to inflation, commodity prices (as outlined in the short term above) and energy prices. The guidance related to the development of life cycle costing attempts to address some of these uncertainties; for example, the framework proposed by Directive 2010/31/EU published by the European Union recommends adopting a constant market interest rate for calculating discount rates, which ignores the possibility of variations over the life cycle of the building resulting from changes in national and international monetary and fiscal policies. It is, nevertheless, clear that changes in price levels due to both the business cycle and inflation are going to erode the accuracy of the cost predication. Another uncertain area in LCC forecasting over the life cycle of the asset relates to determining the service life of building components; this issue feeds back to the central question of data accuracy and reliability. Whilst manufacturer's literature will provide indicative life spans for components, these are often based on factory testing under ideal situations which are rarely encountered in reality. The diesel emissions scandal and Grenfell inquiry have both raised concerns about the accuracy, reliability and, regrettably, the honesty of such factory testing data.

### 7.4.1 Benefits of LCC

Despite the many challenges associated with life cycle costing, the benefits of implementation should not be overlooked. Whilst the case for adopting LCC has been made extensively at the start of this chapter, and thus will not be repeated here, it is nevertheless important to capture some of the core messages. Firstly the PFI/PPP (private–public partnerships) market has demonstrated the benefits of proactively modelling an asset's life cycle costs. Through doing this, it has been possible for investors to make decisions based on value. By looking at the balance between capital and operational costs, decisions can be made as to how the design can be developed.

This can be best illustrated through a case study example, in this case the Princess Margaret Hospital in Swindon. The hospital was procured using a PFI model with the project led by Carillion. In this example, Carillion adopted life cycle costing to inform material selection decisions with a view to achieving both the internal financial performance requirements whilst also satisfying the client's (NHS) output specification which required the inclusion of 'green building principles', including targets related to lifetime savings of 30% in $CO_2$ emissions and 50% in waste generation (Edwards et al. 2000; NHS Estates 2001). In the example, a design decision had to be made in relation to the specification of the roof structure. To achieve these two requirements the contractor looked at, amongst other elements, the design of the roof, or more specifically the levels of insulation to be provided, with the proposal that insulation levels could be doubled. Initially looking at the capital cost, doubling the insulation would cost £21,000; however, the additional insulation would reduce the number of radiators needed on the top floor of the hospital, generating a saving of £27,000 on the capital costs of radiant appliances. Furthermore, the life cycle costing exercise revealed that in addition to the capital saving of £6,000, a further £213,000, it was estimated, would be saved in terms of maintenance and running costs over the asset's life cycle. The total financial benefit from this design change was around £219,000.

## 7.5 Undertaking Life Cycle Costing

The first section of this chapter has looked at the benefits and challenges of life cycle cost analysis whilst also making the case for implementing life cycle cost analysis as part of an effective thorough life asset decision process. Now these issues have been appraised, the second part of the chapter will focus on the process of undertaking a life cycle analysis. This section of the chapter will address a number of issues including the process of calculating the time value of money, how the time horizon for the study is to be determined, determining the life expectancy of components, building in-use considerations and, finally, how the discount rate applied to the time value of money calculations is determined.

### 7.5.1 Time Value of Money

Value for money is an economic concept that refers to anything that involves forecasting, estimating or predicting costs into the future. This is the mathematical aspect of life cycle costing that the majority of people, especially at university, find incredibly confusing. Yet in reality this is a reasonably simple process, but often the formulas used for these calculations terrify and put people off proceeding with the calculations.

In an attempt to aid the reader with these calculations, we will start at the very beginning. At the core of these calculations is the question, which would you prefer, to receive £100 today or £100 in one year? This concept is called the *time preference of money*. For the majority (whom economists call 'rational' people) the answer is always to take the £100 today rather than wait for the same £100 in one year. The reason people prefer not to wait is due to the perceived risk, uncertainty or the reality that £100 today will not be worth £100 in 12 months. This in turn raises a second question, why do people save money in the bank or invest in property? The answer to this is complex, but essentially boils down to the desire to have some financial security, but they also invest in the hope of making money – a return on their investment or compensation for their decision to defer their consumption (HM Treasury 2020).

**Calculating Interest**

Taking this simple concept a little further, let's ignore the return element for a moment and look at someone who wishes to save their money; they do so hoping to make some interest. Interest in this context is really the reward the bank will give its customers to reward them for saving money. This interest can be applied to the money saved in the bank in two ways. The first is via a concept called simple interest. In a nutshell, simple interest is a quick way of calculating interest. In this situation, interest is calculated by taking the percentage interest rate and multiplying it by the number of periods (days, weeks, months or years) the money will be invested for. This process is undertaken using the following formula:

$$Simple\ interest = P * r * n$$

In this formula, P represents the sum of money invested, r is the rate of interest and n is the number of compounding periods. To demonstrate how this formula works, let us take the example of £100 and assume the bank is offering an interest rate of 2% and we are investing the money for two years. The calcuation undertaken in Example 7.1, shows that over the investment period, £4 of interest would be generated. An important point to make here is that in the formula the rate of interest is always provided as a decimal not a percentage. As shown in Example 7.1 to change a percentage to a decimal you simply divide by 100.

**Example 7.1**

$$Simple\ Interest = P * r * n$$

In this example, P will be £100.00, r will be 2% and n will be 2 years.
*Stage One* – Change the percentage to a decimal:

$$Percentage\ to\ decimal = \frac{r}{100}\ so\ ...... \frac{2}{100} = 0.02$$

> *Stage Two* – Calculate the simple interest:
>
> $$Simple\ Interest = £100 * 0.02 * 2$$
>
> $$Simple\ interest = £4.00$$

Although simple interest is not often, if ever, used when determining interest on bank accounts, it is regularly used in construction. For instance, in NEC standard forms of contract, if the employer does not pay the contractor by the due date, they can be requested to pay the contractor simple interest on the outstanding funds. More regularly, a different form of interest, called 'compound interest', is applied to invested money. In this situation, money invested will make interest over the first period, the cumulative of which will then be reinvested and interest in the second period will be added to both the original sum invested and the interest earned in the first year. Compound interest is calculated using the following formula:

$$Compound\ interest = P(1+r)^n$$

In this formula P once again represents the sum invested, r is the rate of interest and n is the number of compounding periods. Taking our earlier example again of £100 invested at a rate of 2% with the money invested for the same two years.

---

**Example 7.2**

$$Compound\ interest = P(1+r)^n$$

In this example, P will be £100.00, r will be 2% and n will be 2 years. Calculate the compound interest:

$$Compound\ Interest = £100(0.02+2)^2$$

$$Compound\ interest = £104.04$$

---

In Example 7.2, you can see the compound interest is £4.04. Whilst this does not show a significant difference between simple and compound interest, the difference is more evident if you look over the longer time period, say, 10 years. With simple interest, the interest received would be £10, whereas compound interest would be £21.90.

### The Cost–Time Relationship

As explained earlier, rational people would rather have the £100 today than wait for the £100 in 1 year, as they know £100 will be worth less in 12 months. It is acknowledged in economics that a sum of money will be worth less in the future than it is today. In some texts this is termed the cost/time relationship. Understanding this concept is essential to the execution of life cycle costing but it is also the concept that most students introduced to life cycle costing for the first time struggle to understand. The cost/time relationship is a very important consideration when undertaking life cycle analysis, as future sums would need to be calculated and would then need to be adjusted back into one period of time to make the analysis meaningful and useful. To achieve these two objectives, two principal techniques are used: the first is 'future value' and the second is 'present value'.

### Future Value

Future value is the natural extension of the concept of compound interest and is concerned with forecasting the value of a sum of money sometime into the future assuming a certain rate of interest will be applied. For instance, if you ever find yourself in the fortunate position to have a reasonable sum of money available but rather than spend it, you want to put it away to replace your car. You can apply the future value of money concept to see what it would be worth. As with interest calculations the concept can be applied to both simple and compound interest using the following two formulas:

$$FV \text{ using simple interest} = P(1 + in)$$

$$FV \text{ using compound interest} = P(1 + r)^n$$

where $P$ is the sum of money invested (also called the principle), $r$ is the rate of interest expressed as a decimal (see above) and $n$ is the time (number of periods or in this instance number of years). These formulas are used in Example 7.3 to illustrate how future value works.

---

**Example 7.3**
A sum of £5,000 is invested for a period of 5 years in a reasonably high interest account paying 3% per annum. What will the value of the investment be at the end of the period?

$$Future\ Value(compound) = P(1 + r)^n$$

$$Future\ Value(compound) = £5000(1 + 0.03)^5$$

$$Future\ Value(compound) = £5,796.37$$

---

The more vigilant of you will have noticed the calculation in Example 7.3 looks very similar to the calculation in Example 7.2; this is because the second formula is identical to the compound interest formula. Just to reassure you, this is not a typographical error. Essentially, future value is doing the same calculation as you would do when computing compound interest. The simple interest formula is again the same as previously; this time it is written more mathematically but the outcome will be the same.

Although future value is not often used in the execution of life cycle cost analysis, understanding the principles of this will help with the next technique, present value.

### Present Value

As explained earlier, rational people would rather have the £100 today than wait for the £100 in 1 year, as they know £100 will be worth less in 12 months. It is acknowledged in economics that a sum of money will be worth less in the future than it is today. We are now at the stage where we can calculate the equivalent value of that money at today's prices. To do this, present value (PV) is used. The RICS (2014, p. 17) defines present value as 'the cost or benefit in the future discounted back to some base date, usually the present day, at a given compound interest rate'. A simpler definition is provided by HM Treasury (2020), which defines present value as 'the future value expressed in present terms by means of discounting'. No doubt this has all suddenly become very technical and very complex. However, in reality this concept is simple. Let's ignore the technical terminology, such as discounting, and revert back to the money for the car in the example above. In Example 7.3 we calculated the future value of £5,000 assuming this has been invested into a bank for 5 years and 3% interest. The outcome of the calculation was FV = £5,796.37. So, what is the present value of that sum? The answer can be seen in Figure 7.3 as the sum is moved back and forwards in time. Very simply, present value represents the amount of investment today required to pay for the capital cost plus all future operating (revenue) costs (RICS 2014).

As can be seen illustrated in Figure 7.3, present value is a means of calculating the current economic value of a sum of money taken from sometime in the future, to enable that sum of money to be compared with other sums of money in the same time period. This is an especially important feature of life cycle costing, given the desire to look at capital and operational costs side by side.

The present value calculation is undertaken using the following formula:

$$PV = \frac{1}{\left(1+r\right)^{n}}$$

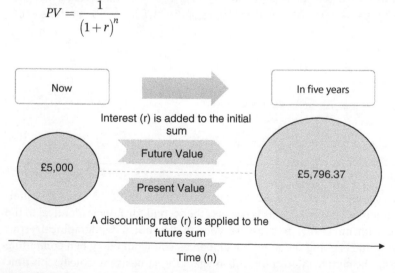

**Figure 7.3** Cost–time relationship with FV and PV indicated.

where $r$ is the discounting rate expressed as a decimal and $n$ is the time period over which the discounting process to be applied. A worked example demonstrating the use of this formula has been provided in Example 7.4.

---

**Example 7.4**

Calculate the PV (present value) of £5,796.37 in 5 years, discounted at 3% per annum.

$$PV = \frac{1}{(1+r)^n}$$

$$PV = \frac{1}{(1+0.03)^5}$$

$PV = £4,999.9996$ *which can be rounded up to* £5,000

---

This concept can now be applied to an example more akin to those you would find in a maintenance environment. Let's assume you are undertaking the life cycle cost analysis for a roof over an 80-year period. The manufacturer's guidance for the roof covering suggests replacement will be needed every 20 years; therefore, a series of PV calculations are going to be required. The approach to dealing with this calculation is illustrated in Example 7.5. In the example it is assumed the roof replacement, including removal of the old roof covering, will cost £20,000 at current prices. A discount rate of 5% has been assumed for this example.

---

**Example 7.5 – Periodic Roof Replacement**

Stage One – Calculate the PVs for £1 at each of the replacement stages; the PV formula has been applied in each row, but only one calculation is fully demonstrated. Calculation for year 40:

$$PV = \frac{1}{(1+r)^n}$$

$$PV = \frac{1}{(1+0.05)^{40}}$$

$$PV = 0.1420$$

Calculate the PV for each replacement cycle and add the total together.

| PV of £1 @ | | |
|---|---|---|
| | Year 20 | 0.3769 |
| | Year 40 | 0.1420 |
| | Year 60 | 0.0535 |
| | $\Sigma$ | 0.5724 |

---

> The cost of replacing the roof is now multiplied by 0.5724 to give the life cycle cost of replacement at today's prices. So that would be:
>
> £20,000 x 05724 = £11,448.00

In the example you will have spotted that the roof has a lifespan of 80 years, but the calculations only consider replacement of the covering in years 20, 40 and 60. The reason year 80 is ignored is that repair, maintenance and replacement costs are normally not allowed for in the final year of a building's life. Given the life of the building is 80 years, and the roof is replaced every 20 years, no replacement will be allowed for in year 80. As a consequence, the last time the roof covering would be replaced is year 60.

If, however, the building was not to be demolished or face some other end-of-life outcome after year 80 this situation would change. Consider, for example, the case of a PPP project. Whilst year 80 may represent the end of the concession period and thus the end of the contractor's life cycle modelling, this is the period the contractor is responsible for. It is likely there will be a contractual obligation to hand over the building with defects rectified and maintenance carried out to a minimum specified standard. As such, it would be expected that the year 80 roof covering replacement works would be completed as part of those obligations.

### Present Value of £1 per Annum or Years Purchased

The second present value calculation you are likely to encounter when undertaking life cycle cost analysis is 'years purchased' or the 'present value of £1 per annum'. The earlier present value calculation assumes costs are incurred periodically for example costs associated with component replacement, such as in the example where the roof was replaced every 20 years. Yet not all maintenance, repair and replacement activities will be undertaken quite so sporadically. In some situations, annual or more frequent interventions would be needed. Whilst these can be effectively modelled using the PV formula introduced earlier, this would involve a lot of calculation and would make the model incredibly complex. To avoid this, where maintenance is occurring as a series of interventions such as annual cleaning of high-level windows, years purchased or the present value of £1 per annum is used.

The present value of £1 per annum or years purchased calculation is undertaken using the following formula:

$$PV = \frac{\left[(1+r)^n - 1\right]}{[r(1+r)^n]}$$

For many this formula is a little unwieldy, although the formula does work within an excel spreadsheet, it is easier to approach the calculation in two stages:

Stage One – Calculate the Present Value

$$PV = \frac{1}{(1+r)^n}$$

where r is the rate of interest (as a decimal) and n is the time period.

Stage Two – Calculate the Years Purchased

$$Years\ Purchased = \frac{(1-PV)}{r}$$

where PV is the outcome of the present value calculation and r is the rate of interest as a decimal.

This concept can now be applied to a scenario more akin to those you would find in a maintenance environment. Let us assume we are undertaking a life cycle cost analysis for windows in a high-rise office building. The study is conducted over a 30-year period. As part of the analysis, annual maintenance is also required to clean the windows using a high reach platform. Example 7.6 assumes this annual maintenance will cost £2000 at current prices. As before, discount rate of 5% has been assumed.

---

**Example 7.6 – Annual High-Level Window Cleaning**

In this example the annual maintenance costs will be calculated for the full 30-year period using the PV of £1 per annum or years purchased formula:

$$PV\ of\ \pounds1\ per\ annum = \frac{\left[(1+r)^n - 1\right]}{[r(1+r)^n]}$$

$$PV\ of\ \pounds1\ per\ annum = \frac{\left[(1+0.05)^{30} - 1\right]}{[0.05(1+r)^{30}]}$$

$$PV\ of\ \pounds1\ per\ annum = 15.372$$

This can now be multiplied by the annual maintenance cost of £500 to give the total life cycle cost, discounted to today's price of £2000 x 15.372 = £30,744 – which is the cost of cleaning at today's prices.

If you compute this using the simpler process, the outcome will be the same, as shown below:

Stage One – Calculate PV of £1.

$$PV = \frac{1}{(1+r)^n}$$

$$PV = \frac{1}{(1+0.05)^{30}} = 0.2313$$

Stage Two – Calculate YP of £1:

$$Years\ Purchased = \frac{(1-PV)}{r}$$

$$Years\ Purchased = \frac{(1-0.2313)}{r}$$

$$Years\ Purchased = 15.3721$$

As you can see, this is the same outcome as that noted above in the more complex formula.

### Annualised Equivalent Value and Sinking Funds

One of the final economic measures to introduce in this section is the concept of annual equivalence. The RICS guidance note on life cycle costing (RICS 2016) defines this as 'the present value of a series of discounted cash flows expressed as a constant annual amount'. It goes on to suggest that it is in effect the annual costs of investing in a building and can be equated to the interest that would have been gained and expected per annum (simple interest not compound interest) had the money been left in the bank. Obviously, such a position assumes there would be no residual value left in the building at the end of its life cycle. For this reason, it is better to see this relationship not with the building but with a component in the building that will need to be periodically replaced.

A good example of this can be taken from the domestic private rental sector, where periodic replacement due to damage is a known liability. Let us assume the investor is replacing an end-of-life kitchen; we will also assume the replacement kitchen will not add anything to the building's value. The replacement kitchen is going to cost the investor £4,000. This will be taken from a bank account paying interest at 2.5% per annum. In this example, there is a theoretical loss of £80 per annum. The problem with investing in this replacement kitchen is that at the end of the kitchen's life, all that is left is waste. So, in reality, the loss to the investor is greater than simply the £80 per annum. The investor is not only losing the interest on the initial capital sum, the £80 per annum they are also losing the money they need to save to replace the kitchen at the end of its life in, say, 10–15 years. This need to save money for the next replacement kitchen is called a sinking fund, this can be calcuated using the annual equivalent value.

The annual equivalent value is calculated using the following formula:

$$AEV = \frac{cr}{(1+r)^n - 1}$$

where C is the cost in year n, r is the discount rate and n is the time (number of years). Example 7.7 demonstrates the use of this calculation, taking the kitchen example as a basis for the calculation.

**Example 7.7 – Annualised Equivalent Value**

In this example, the annualised equivalent value for the replacement kitchen mentioned earlier will be calculated. It is assumed the kitchen will last for 15 years and the discount rate is 2%.

$$AEV = \frac{(cr)}{[(1+r)^n - 1]}$$

$$AEV = \frac{(4000 * 0.02)}{[(1+0.02)^{15} - 1]}$$

$$AEV = 231.3018$$

$$Years\ Purchased = \frac{(1-0.2313)}{r}$$

$$Years\ Purchased = 15.3721$$

Therefore, a cost of £4000 in 15 years' time at an interest rate of 2% will be equivalent to an annual investment of £231.30.

Despite the usefulness of the annual equivalent value calculation, the RICS guidance note on life cycle costing (RICS 2016) suggests its usefulness within the life cycle costing arena is rather limited. Although the annual equivalent can be used to compare options with a consistent annual spend over the time horizon. It is not well suited to the expenditure profile of most buildings which will have some annual expenditure but also period maintenance and replacement cycles. With this in mind, RICS (2016) advocates the use of the discounted cash flow method, based on present value calculations.

**Net Present Value**

The final financial metric used in life cycle costing is the concept of net present value (NPV). Net present value is one of the best-known methods of financial or economic performance appraisal. Defined in the RICS life cycle costing guidance note as the 'total present day worth of a future cash flow discounted at a given interest rate' (RICS 2016), NPV allows the surveyor to model the financial/economic performance of the asset over its full life cycle with various maintenance, repair and replacement activities integrated into the cash flow.

Net present value is represented by the following formula:

$$NPV = \sum_{t=1}^{T} \frac{C_t}{(1+r)^t} \ or \ C_0 = \frac{C_1}{(1+r)^1} + \frac{C_2}{(1+r)^2} + \frac{C_3}{(1+r)^3} + \dots \frac{C_n}{(1+r)^n}$$

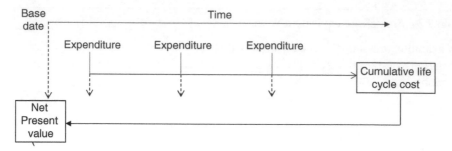

**Figure 7.4** Basic Structure of LCC discounting model *Source*: Adapted from RICS 2020.

This formula presents the mathematical way of describing a cash flow mapped in a spreadsheet with various items of cost reflected over the course of the investment time horizon. This can be shown more clearly in a simple example, this is provided in Example 7.8 and is illustrated in Figure 7.4.

---

**Example 7.8 – Net Present Value**

In this example, a discounted life cycle cost appraisal is presented for rainwater goods. The building has an anticipated life cycle of 25 years. The initial costs are £7,500 for seamless aluminium guttering and rainwater pipes and £4,250 for uPVC gutters and rainwater pipes, the aluminium is expected to last for 20 years and the uPVC for 12 years. Replacement costs are equal to initial installation costs with a 10% allowance for removal and disposal. A discount rate of 3.5% is recommended.

**Option One – Aluminium**

| Cost | Cost (£) | Year | PV of £1 | PV (£) |
|------|----------|------|----------|--------|
| Initial cost | 7500 | 0 | - | 7500.00 |
| Replacement | 8250 | 20 | 0.3768 | 3108.60 |
| | **Discounted life cycle cost** | | | **10,608.60** |

**Option Two – uPVC**

| Cost | Cost (£) | Year | PV of £1 | PV (£) |
|------|----------|------|----------|--------|
| Initial cost | 4250 | 0 | - | 4250.00 |
| Replacement | | 12 | 0.5568 | |
| | | 24 | 0.3100 | |
| | 4675 | Σ | 0.8668 | 4052.29 |
| | **Discounted life cycle cost** | | | **8,302.29** |

The example above demonstrates the application of discounted cash flow or net present value. In the examples, the individual PV of £1 have been determined using the PV calculation outlined earlier. In each case the replacement cost is then multiplied by this value.

As you can see from the example, despite having two replacement cycles uPVC offers a cheap whole-life cost.

## 7.5.2    Determining the Time Period of Appraisal

Now the process of calculation has been explored, the next decision facing the asset management surveyor, quantity surveyor or other professional relates to the determination of the time horizon over which the study will be executed. For some situations, this will have already been determined, for instance in a PFI or PPP situation this would be the length of the concession period. In most other situations the client is the one who will determine the study duration but they are likely to seek advice on this from their professional team.

To determine the preferred time horizon, it is important to determine the operating, design and functional life of the asset. These terms are important, and they mean very different things, defining them is therefore important:

- *Design Life* – The expected period the designers expect the building to last based on the lifespan of the main components.
- *Operating Life* – The actual lifespan the building attains, can be longer or shorter than the design life, depending on maintenance, upkeep, refurbishment or rehabilitation.
- *Functional Life* – The time span before the building will cease to function in its current use; usually this is related to the problem of obsolescence.

Whilst it is to an extent possible for surveyors to analyse data relating to when buildings are typically demolished to predict the asset's operating life, the functional life of buildings is usually far shorter than the true operating life of the asset. In general, buildings fail for two major reasons as illustrated in Figure 7.5. These are:

- Deterioration resulting from age and component failure;
- Obsolescence or changes not related to the structure, but its functionality.

It is important therefore that a distinction is drawn between these two primary causes of building failure. Deterioration is largely due to the age and use of the building, although as Table 7.2 shows there are other common causes of failure. As buildings age they naturally start to develop defects, these can usually be corrected through routine maintenance. By the same token, as parts of the building wear out, such as the windows, they will be replaced to preserve the building's usability and functionality. The levels and speed of deterioration can be to an extent controlled during the design of the building or refurbishment scheme. For instance, the designer can select products with a longer or shorter life cycle or the team can consider likely usage levels and source a product offering higher levels of durability but perhaps at a higher capital cost. Equally, when managing the asset during occupancy, components can be controlled by proactively managing their

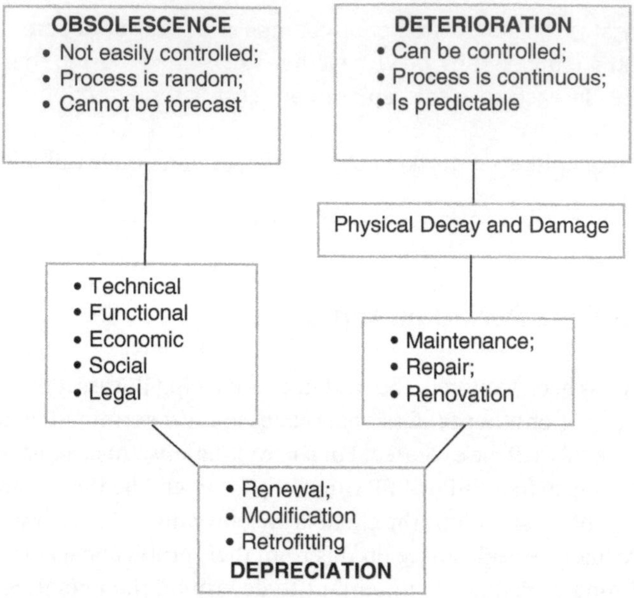

**Figure 7.5** Obsolescence, deterioration and depreciation. Adapted from Flanagan 1989 and RICS 2018.

**Table 7.2** Causes of building deterioration.

| Main Factors | Causes of deterioration |
| --- | --- |
| Weathering Factors | Radiation (solar, infrared and ultraviolet) |
| | Temperate |
| | Water |
| | Other chemical agents such as sulphates |
| | Normal air constituents |
| | Freeze–thaw |
| | Wind |
| Biological Factors | Insect vermin |
| | Surface growths (Algae, bacteria, fungi) |
| | Animal vermin |
| | Plant agents (tree routes, ivy) |
| Stress Factors | |
| Incompatibility Factors | Chemical |
| | Physical |
| Use Factors | Manufacture and Construction |
| | Design |
| | Building occupants and other users |
| | Maintenance |
| Other Natural Factors | Ground Movement (soil movements) |

Adapted from Mydin et al. (2012).

replacement cycle through preventive maintenance planning, as explained elsewhere in this book. Returning back to our carpet example, the building manager/asset manager could look to replace floor coverings in areas of heavy wear at a time that is convenient for the organisation, rather than waiting for the carpet to reach the end of its life and undertake reactive maintenance, thereby maintaining the asset's functionality and managing the client's financial exposure.

Obsolescence, on the other hand, is much harder to control, as it is triggered by unforeseen events. Changes in fashion, technology or social systems, for instance, that consequently affect the design and use of buildings are clear examples of obsolescence. Obsolescence and the effects of these unforeseen events can be witnessed through the demise of the office buildings. These structures are now being demolished, in some cases after as little as 20 years following their initial construction, despite the structure having not reached the end of its natural life (Pinder and Wilkinson 2001). In the main, this has been due to changes in occupant demand rendering such structures functionally obsolete. For example, the use of IT and the need for larger internal floor-to-ceiling heights to allow raised floors and ceilings to be installed to support installations such as CAT5 network cables and air handling requirements. Deterioration in a building can ordinarily be remedied at a cost to the building owner, assuming repairs are economically viable when compared to replacement of the asset. Obsolescence is much harder to remedy. In some situations the building will not accept the levels of change required to counter the effects of obsolescence and will need to be demolished as in the case of the office buildings outlined earlier. The main categories of obsolescence are explained in Table 7.3.

Looking at the practical application of physical failure and obsolescence in the built environment, you would expect a house to last well over 100 years, although the design life for most modern houses is only 60 years. However, as very little changes with the way we live, houses can be refurbished and upgraded to keep them functioning as the owner desires with relative certainty that the economic analysis will always favour refurbishment as property desirability will enhance value. Yet, by the same token, the high-rise tower blocks constructed during the 1960s have mostly been demolished. This was not because the way people live have changed, but as a result of a combination of deterioration due to poor design and low build quality, alongside socio-economic obsolescence. Most tower blocks became blighted by anti-social behaviour leading to unpopularity and low demand amongst social housing tenants. As a result, they became socially obsolete. It must also be remembered at this point that some buildings are demolished neither due to deterioration or obsolescence; they are demolished because the land they occupy has more value in an alternative use or they are simply in the way of major infrastructure developments.

As a result, the asset management professional needs to forecast the operative life of the building whilst trying to foretell the social, technical and other changes. For this reason, life cycle cost studies are typically undertaken based on duration of the owner's interest in the building, so for a PPP project, this would be the length of the concession, whereas for some grant funders in the public sector, such as Sport England, a 20-year analysis will often suffice.

**Table 7.3** Categories, definitions and basis of assessment for obsolescence.

| Type of Obsolescence | Definition | Basis for assessment of building life | Examples of factors leading to obsolescence |
|---|---|---|---|
| Physical (Deterioration) | Life of the building to when physical collapse is possible | How long will the building meet human desires (with the exclusion of economic considerations) | Deterioration of the external brickwork<br><br>Deterioration of the concrete frame |
| Economic | Life of the building to when occupation is not considered to be the least cost alternative of meeting a particular objective | How long will the building be economic for the employer to own or operate? | The value of land on which the building stands is more that the capitalised full rental value that could be recovered from the building |
| Functional | Life of the building to when the building ceases to function for the same purpose as it was constructed | How long will the building be used for the purpose for which it was built? | Train stations converted into industrial, retail and residential dwellings<br><br>Cinemas converted into bingo halls<br><br>Old mills converted into prestigious apartment buildings in city centres |
| Technological | Life of the building until the building is no longer technologically superior to alternatives | How long will the building be technologically superior to alternatives? | Prestige offices unable to accommodate introduction of high levels of computing facilities<br><br>Storage warehouse unable to accommodate the introduction of robotics in goods handling |
| Social | Changes in the needs of society result in a lack of use for certain buildings | How long will the building meet with social desire of the population (with the exclusion of economic considerations)? | Churches converted to restaurants, retail units and residential dwellings<br><br>1960s high-rise tower blocks becoming sink estates leading to high levels of anti-social behaviour |
| Legal | Legislation resulting in the prohibitive use of buildings unless major changes are made | How long will be building meet legislative requirements? | Major changes to buildings following the Introduction of asbestos controls |
| Aesthetic | Style of architecture no longer fashionable | Brutalist architecture of the 1960s/1970s | Buildings clad with concrete or ceramic tiles; 1960s office buildings |

Adapted from RICS (2014).

## 7.5.3   Component Life Considerations

Once the asset management professional has considered the overall design life of the building, the surveyor needs to consider the lifespan of the individual components. Given a building consists of a large number of individual components, all with different maintenance requirements and design lifespans, this is not an easy task, especially as trade literature will often provide vastly different life expectancies for similar materials. Added to this is the complication that the majority of this data is based on controlled experiments and will not necessarily reflect the realities of the building during its use phase.

It is, however, possible to access real component life data as part of a BCIS subscription. The data included in component life represents the findings of regular surveys issued to a sample of 80 building surveyors, who are asked to report on the life expectancies for a predetermined list of common building components. This data is subsequently analysed ansd made available via the BCIS subscription service.

However, it has to be remembered that all data relating to components' durability will be affected by a range of building specific factors including, but not limited to:

- Level of specification of materials
- Quality of initial installation
- Interaction with other materials
- Levels of use and abuse
- Frequency and standards of maintenance
- Use (heavy or light)
- Local conditions (weathering for instance)
- Acceptable level of performance by the user.

It is, however, possible for the asset manager to adjust typical life expectancy data, by considering the impact of these factors, the process to be adopted is clearly outlined in NRM3 (RICS 2014, p. 90). To aid the reader, this process has also been illustrated in the example below, which considers the life expectancy for a tiled roof installed on a library building. As you can see from the analysis in Example 7.9, when the specifics of the project are considered, the lifespan of the component is significantly reduced.

---

**Example 7.9 – Adjusting Life Cycle Cost Data: Library Roof**

Your client, Authors Borough Council is looking to install a new roof to a small community library facility in Cumbria as part of a planned maintenance scheme. To inform future maintenance planning the local authority have requested a life cycle analysis be undertaken for the roof to determine the possible lifespan of the covering.

- The proposed roof tiles are a flat concrete tile sourced from a leading manufacturer.
- The design will be completed by in-house building surveyors.
- The work will be completed by a competent contractor selected from a framework.

- The location in Cumbria is inland but due to its height above sea level it is exposed to extreme weather in the winter.
- The roof will be regularly maintained.

Using this information and the BCIS data (the median value) we can adjust the life expectancy by comparing the scenario to the reference data (assumed) from BCIS.

| Factors | Reference service life assumptions | Project conditions | Factors |
|---|---|---|---|
| Quality of components | Average | Standard | 1 |
| Design level | Not known | Good | 1.1 |
| Work standard | Good practice | Good practice | 1 |
| Exterior environment | UK average | High exposure | 0.8 |
| In-use conditions | External envelope | External envelope | 1 |
| Maintenance level | Manufacturers recommendations | Maintained regularly | 1 |
| Overall factor | 1.00 | | 0.88* |
| *Reference service life* | *63 Years* | *Project Service life* | *55.44 Years* |

*To calculate multiply the reference factor of 1.00 by the factors determined.

Analysis at component level is important to clients with large property portfolios, as they will look to translate the life cycle analysis into a maintenance plan for the building with a view to controlling their exposure to component failure risk. For this reason, most major organisations proactively manage components so they are replaced under controlled conditions before they fail, therefore requiring reactive maintenance. Reactive maintenance is financially difficult to control and impacts negatively on the building users, leading to potential disputes. Similarly, the owners of large property portfolios will routinely undertake periodic inspections of between 5 and 10% of their stock to allow them to plan their maintenance investment. As a result, the asset manager maybe tasked not with predicting the component's life expectancy, but the length of time the component will be retained.

## 7.5.4 Discount Rate, Interest Rate and Inflation

Life cycle cost planning differs massively from the approach to cost planning discussed in Chapter 6, mainly due to the timeframes involved. Cost planning forecasts the construction costs for the building, usually focused on the final account that will be agreed shortly after practical completion. Life cycle costing needs to consider the buildings occupancy, running and maintenance costs, so the asset manager is now trying to predict cost from practical completion, for a period of 20 to 60 years. As a result, the asset manager needs to consider both current costs and future costs.

The difficulty with future costs is the impact of *time value*, the effects of risk over and above a risk-free investment, inflation and interest. As a result, the asset manager cannot simply add numbers together that represent different phases of expenditure. The asset manager needs to use a method of calculation termed the net present value (NPV). Net present value was discussed earlier in this chapter. To calculate the NPV for each year in the life cycle analysis, the asset manager needs to identify an appropriate discount rate, which ensures the time value of money is taken into account. It is important that you do not mistake the *discount rate* for the *interest rate*. To clarify the key differences, both these concepts are explored.

### Discount Rate versus Interest Rate

Earlier in this chapter, these two concepts are used interchangeably; however, it is critically important that these concepts are well understood. Interest rates by their virtue are an easier concept to understand. The interest rate is the return on investment paid by banking institutions to reward investors for depositing funds with them. Equally, interest can also be charged by financial institutions against those who are seeking to borrow funds. For example, via mortgages, business loans, etc.

The concept of a discount rate is rather different and there are a range of positions outlined in the literature on how this concept is to be understood. The RICS guidance note for life cycle costing (RICS 2016) defines the discount rate as 'the interest rate used to bring future costs to a comparable time base (base date)', whereas the RICS guidance note for discounted cash flow (RICS 2011) acknowledges that the discount rate will be determined based on 'the investor's perception of risk', suggesting interest rates and discount rates are not aligned.

As noted by the RICS, for some clients the discount rate is seen as the rate of return on their investment and that rate of return is required to compensate them for both the opportunity cost (loss of interest on their money) as well as the risk the investment presents. This can easily be computed by examining an accepted risk-free benchmark, such as government debt bonds (gilts), then adding to this the required allowances for risk; this would be the risk the general market presents, the construction risk premium and specific risks the client feels they need to add in response to the asset.

For the surveyor selecting the discount rate, ISO 15686-5 and the NRM3 (RICS 2014) guide the surveyor to consider selecting one of the following:

- The opportunity cost of capital
- The clients selected discount rate
- The societal rate of time preference (for public sector projects)
- The cost of borrowing the funds
- Returns lost on investments elsewhere (bonds or equities, etc.)
- The cost of generating equity
- The WACC Rate (weighted average cost of capital).

For most private sector projects, the asset manager will normally adopt the client's developed or target discount rate developed by others as part of the initial investment appraisal. However, if you are developing the life cycle cost plan for a public or third-sector client, it is likely the societal rate of time preference would be used, published by HM Treasury in the *Green Book*, (2020) with the current rate being 3.5%.

The societal rate of time preference, as defined in the NRM3 (RICS 2014, p. 505), is a discount rate that reflects a government's judgement about the relative value which society as a whole assigns (or which the government feels it ought to assign) to present versus future consumption. The societal time preference rate is not observed in the market and bears no relation to the rates of return in the private sector, which are developed based on individual investors' perceptions of consumption, risk preference and investment preference.

The final consideration the asset manager faces is whether to include or exclude inflation; in other words, should the surveyor select *nominal* or *real* discount rate? In their advice on LCC, the RICS guidance note recommends the use of real costs (those relevant at the time of analysis). They further suggest 'there is a typical assumption that inflation and/or deflation rates apply equally to all costs and therefore can be ignored' (RICS 2016, p. 3). If, on the other hand, nominal rates (those estimated at the time of expenditure) are used the guidance note suggests they 'will have been adjusted for inflation/deflation and estimated efficiency or technological change' (RICS 2016, p. 3).

As a rule, LCC studies normally work on the assumption that inflation is an impossible economic variable to forecast to anything like an acceptable degree of accuracy, as a result life cycle cost analysis, and the discount rates selected tend not to include any provision for inflation. At the core of this argument is the assertion that there is often only a small change in the relative values of the various items within the life cycle cost plan. Thus, a future increase in the values of the cost of building components is likely to be matched by a similar increase in terms of other good and services (Ashworth and Perera 2015). This is further complicated by the significant range of building components as a result price changes due to inflation are not uniform and will not necessarily increase with inflation.

The final point on the selection of the discount rate is to consider the risk this presents, when the effects of variations in discount rates are considered the effects on the present value of components can be clearly seen. Figure 7.6 illustrates the different effects of three

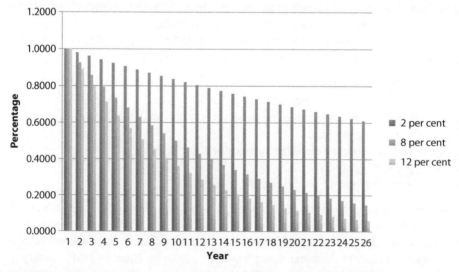

**Figure 7.6** Effects of different discount rates. Adapted from Higham et al. 2017.

**Table 7.4** Discounting data.

| Discount Rate | Year 5 | Year 10 | Year 15 | Year 20 | Year 25 |
|---|---|---|---|---|---|
| 2 % | 0.9057 | 0.8203 | 0.7430 | 0.6730 | 0.6095 |
| 3 % | 0.8626 | 0.7441 | 0.6419 | 0.5537 | 0.4776 |
| 8 % | 0.6806 | 0.4632 | 0.3152 | 0.2145 | 0.1460 |
| 12 % | 0.5674 | 0.3220 | 0.1827 | 0.1037 | 0.0588 |

different discount rates, 2%, 8% and 12% on present value costs. As you can see from the data, the higher the discount rate and the longer into the future you predict, the less variability there is to the whole-life cost.

Table 7.4 illustrates the effects of this by using the three percentage values (2%, 8% and 12%) to calculate the present value for a tile roof covering (150 m$^2$) with a capital cost of £21,750. For example, the table suggests that in year 5, using a discount rate of 2% would yield a present value of £19,698.98 (£21,750 x 0.9057). However, if this was increased to 3%, the value inserted into the life cycle cost analysis would change significantly to a present value of £18,761.55 (£21,750 x 0.8626). That simple adjustment has reduced the component replacement cost by £937.43. Similarly, if the life expectancy of the roof increased to, say, 25 years, with a discount rate of 8%, then the present value would be worth only £1,278.90 (£21,750 x 0.0588). It is, therefore, important that the asset manager applies sensitivity testing to the outcome of the life cycle cost analysis to ensure the effects of change are considered.

## 7.5.5  Building In-Use Considerations

Despite the surveyor taking great care to select the most appropriate cost data, analysis and adjust the life cycle of components, model in the client's proactive maintenance cycles and the select the most appropriate discount rate, life cycle cost modelling can and will be massively affected by the building's end user and the way the building is used. If the building is used and maintained as the surveyor assumes, the life cycle analysis will provide a reasonably close resemblance to the client's maintenance cost profiles; however, this is seldom the situation. It is often necessary for the client's asset management team to replace the components of a building several times during its life, often more regularly than predicted in the life cycle analysis. BCIS (2006, p. 1) advocates, 'All buildings deteriorate naturally but deterioration may be accelerated as a result of human endeavours. It is the human factor that makes determining component and material life expectancy difficult.'

The life of the building as a whole may depend on whether or not efficient and regular repairs and maintenance are undertaken. It not expected this statement will come as a surprise; however, maintenance is often seen as an inefficient use of funds when other financial pressures are applied. Consider for example, the head teacher of a school who has felt austerity reducing the funds available to provide education. Yet the pressures to return GCSE grades remain in order to maintain or enhance league table position remains for that head teacher, who does not have the benefit of a technical construction education. As a consequence, reducing the proactive maintenance budget to increase spending on education (something they are trained to understand and where the pressure is exerted) is the likely outcome. As a result of reduced maintenance, both the life cycle costing model and the life expectancy of the asset are compromised.

It is, therefore, important to remember that life cycle cost analysis tends to reflect the ideal situation or the situation experience at the time the study was produced. However, as with life in general, what happens during the future life of the building is impossible to predict with any level of certainty. Unexpected situations can and will arise which cannot be accounted for within the life cycle costing forecast prepared during the project's design phase.

### 7.5.6   Life Cycle Costing – Applications through the Building Life Cycle

Life cycle costing can be undertaken at various points in the building's life cycle. Ideally the study will be undertaken early, ideally during the project appraisal or feasibility and option evaluation phase, to ensure maximum value of gained from considering potential amendments to both the design of the building and specification of components to reduce the client's exposure to maintenance costs. Or, if that is not possible, to smooth and level that exposure, by trying to design out peaks in component replacement expenditure. Although this is something of an ideal and maybe theoretical proposition, it is still possible to make amendments during the early development of the project to deal with future maintenance, for example, by enhancing a building's *buildability* it maybe possible to reduce maintenance disturbance.

As the project moves into the design stages, further opportunities for life cycle costing will emerge, possibly as part of an overall value engineering exercise or whilst the architect is considering different design features or specifications for major components. For example, in the hotel example used in earlier chapters of the book, a life cycle costing exercise could be conducted to evaluate the whole-life benefits of pod bathrooms and off-site manufacturing as compared to a traditional bathroom installation. Whilst the capital costs maybe prohibitive, value for money could be obtained through lower maintenance and operational expenditure. As with pre-contract cost control, discussed in Chapter 6, cost limits at elemental or sub-elemental level could be determined which take into consideration future costs, thus ensuring the client receives the best possible value for money from their project.

As the project moves through the construction phase, and nears completion, or as part of the government soft landings initiative if public sector, a life cycle cost analysis could be commissioned to prepare a cost-in-use (revenue) budget for the building and to start the phased handover process, by transforming the life cycle cost analysis into an initial asset management and maintenance plan for the asset. In this way, the initial maintenance and asset management plan is informed directly by the design choices and construction methods adopted. This could then be inserted into a level 3 BIM (Building

Information Modelling) model to further centralise and digitalise the main records related to the building. As a result of this, the client will then be able to receive advice throughout the building's life cycle by reconciling actual costs and maintenance interventions against those forecast. For an organisation with a large property portfolio, like a supermarket chain or government organisation that regularly develops a certain building type, this information could be used to further refine future buildings, thereby improving the efficiency of the building's occupancy and maintenance costs.

## 7.5.7 Developing a Life Cycle Cost Plan

Chapter 6 provides a comprehensive discussion of cost planning as it relates to the capital costs of a building. For the vast majority of clients, this will be the service they commission from their quantity surveyor. However, those wishing to adopt a more strategic outlook, will further commission the quantity surveying practice to undertake and produce whole-life cost plans which forecast expenditure for a construction project over its entire life cycle, or a predetermined number of years into the future. With this in mind, the RICS published the New Rules of Measurement in three volumes:

- NRM1 – Order of cost estimating and elemental cost planning – looks at pre-construction cost forecasting;
- NRM2 – Detailed measurement for building works – replaces the SMM7, and guides the surveyor in producing bills of quantities;
- NRM3 – Order of Cost estimating and cost planning for building maintenance works – supplements the NRM1, extending the scope of the document to consider the building's in-use phase.

The NRM3, therefore, represents and extension of the NRM1 and starts to look at helping the client to forecast the costs of maintaining and running the building, by providing maintenance advise to the client. For ease of use, the NRM3 follows the exact same format as the NRM1; however, the NRM3 make use of net present value (NPV) and payback periods to model the short-, medium- and long-term costs of the building, with the purpose of developing both life cycle cost and forward maintenance plans for the asset. This linkage between NRM Vol.1 and Vol.3 is illustrated in Table 7.5, which shows that both documents include formal cost planning phases, but NRM3 requires the quantity surveyor to extend the scope of coverage within the cost plan to include renewal and maintenance costs for the asset.

As Table 7.5 illustrates, NRM3 includes provision a fourth cost planning phase, formal cost plan 4. For a new building this will be an update of formal cost plan 3 and will be based on as-built information, thus taking account of any variations issued during the construction stage. You will have also noticed that the NRM3 expands the required coverage of the earlier formal cost planning phases. As explained in Chapter 6, NRM1 requires the quantity surveyor to produce construct (capital expenditure) cost plans. NRM3 aligns with these and extends the scope of the cost plan to now include not only forecast the capital costs for the building, but to also include forecasts of costs for life cycle renewal works (replacement of elements, sub-elements and components) and annualised maintenance works.

**Table 7.5** NRM Vol. 1 versus NRM Vol. 3 formal cost planning stages.

| RIBA Stage | NRM1 – Pre-contract cost advice | NRM3 – Maintenance cost advice |
|---|---|---|
| 0. Strategic Definition | | |
| 1. Preparation and Brief | Order of cost estimate:<br>• Floor area method<br>• Functional Unit Method<br>• Elemental Method | Order of cost estimate:<br>• Floor area method<br>• Functional Unit Method<br>• Elemental Method |
| 2. Concept Design | formal cost plan 1 | formal cost plan 1 (Renewal/Maintain) |
| 3. Developed Design | formal cost plan 2 | formal cost plan 2 (Renewal/Maintain) |
| 4. Technical Design | formal cost plan 3 | formal cost plan 3 (Renewal/Maintain) |
| 5. Construction | | |
| 6. Handover and Close Out | | formal cost plan 4 |
| 7. Building In-Use | | (Renewal/Maintain) |

To explain this further, the NRM3 (RICS 2014, p. 23) adopts the acronym 'CROME' to describe the key constituents of the life cycle costs of a building and demonstrate broadly how these constituents (CROME) relate to construction costs and to other building maintenance costs. It is also important to point out here that NRM3 does not aim to consider the full range of factors a traditional life cycle cost model includes; it only focuses on renewal and maintenance costs.

CROME stands for:

- C = Construction costs
- R = Renewal costs
- O = Operation and Occupancy costs (not covered by NRM3)
- M = Maintenance costs
- E = Environmental and/or end-of-life costs (not covered by NRM3)

NRM3 (RICS 2014) provides guidance on what costs should be included in each of the letters in the acronym CROME; this is illustrated in Figure 7.7. Unfortunately, due to the size of a complete whole-life cost plan it is not possible provide a fully worked example here. However, guidance on this can be found in NRM3 at pages 67–68, where the basic framework is illustrated along with detailed guidance notes. Should you read this section of NRM3, you will notice that each year is full costed, with renewal costs separated from maintenance costs. All costs are provided at the current time and a discount rate is them applied within the cost plan to identify the NPV of each stream of expenditure. Within the cost plan the costs should be sourced as follows:

- *Renewal Costs* – These will be the capital costs of construction, then subsequent replacements of the component (e.g. carpet) at its capital cost or refurbishment to major

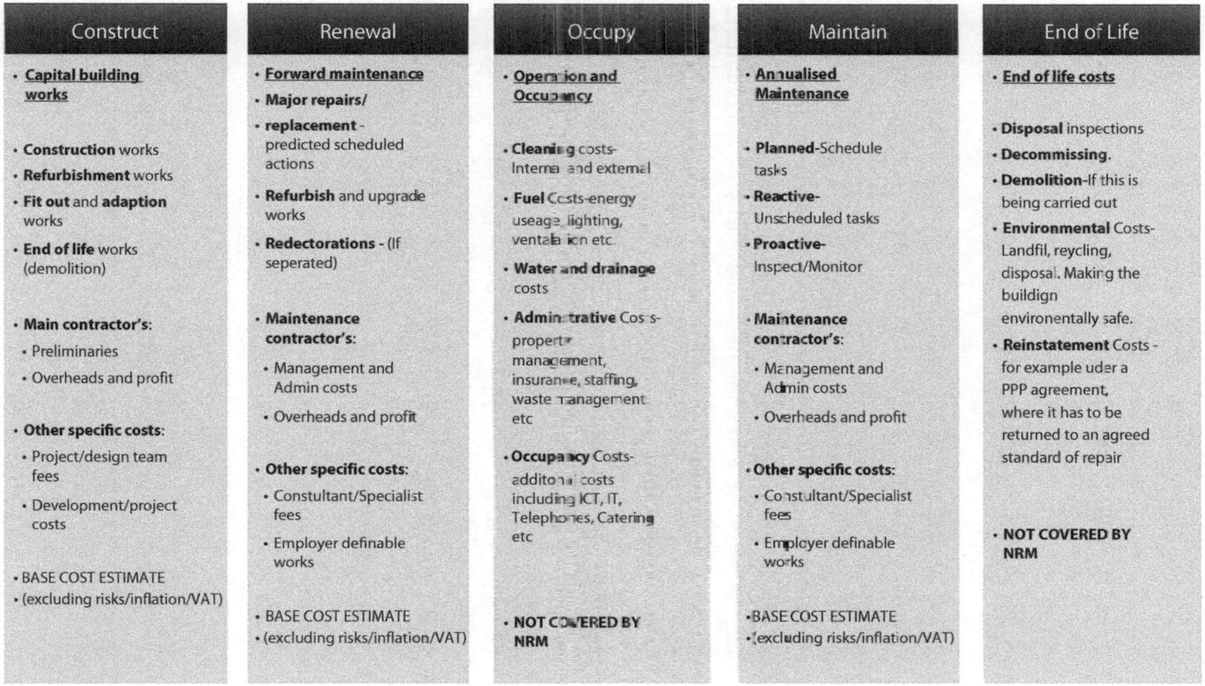

| Construct | Renewal | Occupy | Maintain | End of Life |
|---|---|---|---|---|
| • **Capital building works** | • **Forward maintenance** | • **Operation and Occupancy** | • **Annualised Maintenance** | • **End of life costs** |
| • **Construction** works | • **Major repairs/ replacement**- predicted scheduled actions | • **Cleaning** costs- Internal and external | • **Planned**-Schedule tasks | • **Disposal** inspections |
| • **Refurbishment** works | | • **Fuel** Costs-energy useage lighting, ventaliation etc | • **Reactive**- Unscheduled tasks | • **Decommissing**. |
| • **Fit out** and **adaption** works | • **Refurbish** and upgrade works | • **Water and drainage** costs | • **Proactive**- Inspect/Monitor | • **Demolition**-If this is being carried out |
| • **End of life** works (demolition) | • **Redectorations** - (If seperated) | • **Administrative** Costs- property management, insurance, staffing, waste management etc | | • **Environmental** Costs- Landfil, reycling, disposal. Making the buildign environentally safe. |
| • **Main contractor's:** | • **Maintenance contractor's:** | | • **Maintenance contractor's:** | • **Reinstatement** Costs - for example uder a PPP agreement, where it has to be returned to an agreed standard of repair |
| • Preliminaries | • Management and Admin costs | | • Management and Admin costs | |
| • Overheads and profit | • Overheads and profit | | • Overheads and profit | |
| • **Other specific costs:** | • **Other specific costs:** | • **Occupancy** Costs- additon costs including ICT, IT, Telephones, Catering etc | • **Other specific costs:** | |
| • Project/design team fees | • Consultant/Specialist fees | | • Consultant/Specialist fees | |
| • Development/project costs | • Employer definable works | | • Employer definable works | |
| | | • **NOT COVERED BY NRM** | | • **NOT COVERED BY NRM** |
| • BASE COST ESTIMATE (excluding risks/inflation/VAT) | • BASE COST ESTIMATE (excluding risks/inflation/VAT) | | • BASE COST ESTIMATE (excluding risks/inflation/VAT) | |

**Figure 7.7** Costs included as part of CROME acronym. Adapted from Higham et al. (2017).

elements of the structure (e.g. roof) at their capital cost). So, for example, you may spend £170,000 re-roofing the building in year 50.

- *Maintenance Costs* – The annual expenditure on maintenance should be included. This information can be sourced from BCIS; if you select the 'life cycle costs tab', the system will provide detailed life cycle cost models showing the maintenance breakdown for the building type you wish to analyse. Other sources of maintenance cost data include Building Engineering Services Association *SFG20 Maintenance Task Specifications* and CIBSE *Guide M – 2014 revision*.

## 7.6 Example Life Cycle Cost Models

One of the difficulties of using life cycle cost analysis in practice is the mathematics associated with the evaluation and the time value of money element associated with the need to calculate both present and future values for elements, sub-elements and components.

Whilst the earlier sections of this chapter have sought to provide an overview of the different calculations and explains the theory behind their use, this section of the chapter aims to now demonstrate how these economic principles can be applied and used in the context of a life cycle analysis. The following two examples seek to demonstrate typical scenarios to those that could be encountered when producing life cycle cost plans.

Example 1 – Alternative building designs

The architect has provided the client with two potential designs for their new office building. Both buildings have more or less the same design, but one is designed with a low specification and one is designed with a medium level specification. This is reflected in the capital cost values. The low specification building will cost the client £400,000 whilst the medium specification building will cost the client £550,000. However, it is anticipated that the medium specification building will require significantly less maintenance. The full details of the building are provided in Table 7.6. Based on the information, the life cycle cost of the building can be ascertained, the calculations are illustrated in Table 7.7.

**Table 7.6** Life cycle data for a low and medium specification building.

|  | Low Spec Building | Medium Spec Building |
| --- | --- | --- |
| Initial cost | £400,000 | £650,000 |
| Annual maintenance | £10,000 | £6,000 |
| Quinquennial maintenance | £14,000 | £10,000 |
| Cost of roof replacement | £40,000 | £30,000 |
| Life of roof | 15 years | 20 years |
| Life of building | 50 years | 50 years |
| Discount rate | 10% | 10% |

**Table 7.7** Life cycle cost analysis for low and medium specification levels.

| | Cost Heading | r(%) | n (Yrs) | | Low Spec Building | Medium Spec Building |
|---|---|---|---|---|---|---|
| | **Initial Cost** | | | | £400,000 | £650,000 |
| | **Annual Maintenance** | | | | £99,148 | £59,481 |
| LS | £10,000 x PV of £1 PA | 10% | 50 | 9.9148 | | |
| MS | £6, 000 x PV of £1 PA | 10% | 50 | 9.9148 | | |
| | **5 yearly Maintenance** | | | | | |
| LS | £14,000 x PV of £1 | | | | | |
| | Year 5 | 10% | 5 | 0.6209 | | |
| | Year 10 | 10% | 10 | 0.3855 | | |
| | Year 15 | 10% | 15 | 0.2394 | | |
| | Year 20 | 10% | 20 | 0.1486 | | |
| | Year 25 | 10% | 25 | 0.0923 | | |
| | Year 30 | 10% | 30 | 0.0573 | | |
| | Year 35 | 10% | 35 | 0.0356 | | |
| | Year 40 | 10% | 40 | 0.0221 | | |
| | Year 45 | 10% | 45 | 0.0137 | | |
| | £14,000 | @ | Σ | 1.6155 | £22,617 | |
| MS | £10,000 x PV of £1 | | | | | |
| | Year 5 | 10% | 5 | 0.6209 | | |
| | Year 10 | 10% | 10 | 0.3855 | | |
| | Year 15 | 10% | 15 | 0.2394 | | |
| | Year 20 | 10% | 20 | 0.1486 | | |
| | Year 25 | 10% | 25 | 0.0923 | | |
| | Year 30 | 10% | 30 | 0.0573 | | |
| | Year 35 | 10% | 35 | 0.0356 | | |
| | Year 40 | 10% | 40 | 0.0221 | | |
| | Year 45 | 10% | 45 | 0.0137 | | |
| | £10,000 | @ | Σ | 1.6155 | | £16,155 |
| | **Roof Replacement** | | | | | |
| LS | £40,000 x PV of £1 | | | | | |
| | Year 15 | 10% | 5 | 0.2394 | | |
| | Year 30 | 10% | 10 | 0.0573 | | |
| | Year 45 | 10% | 45 | 0.0137 | | |
| | £40,000 | @ | Σ | 0.3104 | £12,417 | |
| MS | £30,000 x PV of £1 | | | | | |
| | Year 20 | 10% | 20 | 0.1486 | | |
| | Year 40 | 10% | 40 | 0.0221 | | |
| | £30,000 | @ | Σ | 0.1707 | | £5,121 |
| | **Results of the Life Cycle Cost Analysis** | | | | **£534,182** | **£730,757** |

In the calculations in Table 7.7, the sums from Table 7.6 have been reduced to a common timescale by calculating the net present value for each item. In this example, valuation tables have been used for simplicity to determine the present value for each element. This calculation can also be undertaken using a calculator or excel spreadsheet. To do this you would simply use the PV of £1 equation, whilst the 5-year (quinquennial) maintenance is discounted using the PV of £1 formula. Both these techniques are introduced in section 7.4.1.

In the example it can be seen that the initial capital costs have not been adjusted. These are referred to as year zero costs. As they are encountered at the time of the analysis they are assumed to be at the current prices and accordingly do not require any further discounting.

Whilst the example is simplistic, and does not meet the rigours required for an NRM compliant study, the results of the analysis nonetheless suggest that the lower specification will present the client with better overall value for money; however, the analysis does not consider potential differences in rental return, nor does it consider occupancy costs, which may differ significantly. Finally, the analysis does not allow for sensitivities, such as changes in interest rate, target return rate or life expectancy and maintenance levels that will be achieved in practice.

Example 2 – Alternative Window Options

This exercise considers a choice between window specifications. As this building is in a conservation area, only timber sliding sash windows are permitted. In this situation the architect has provided the client with a choice between cheaper redwood sliding sash windows that will be painted with primer, undercoat and gloss, and more expensive sapele (hardwood) windows which will receive two costs of wood stain. Both windows will include 20 mm double glazed units and will achieve the same u-value. Full details for each option are provided in Table 7.8. Based on the information, the life cycle cost of the 10 windows over a 20-year period has been calculated, as illustrated in Table 7.9.

In the calculations in Table 7.9, the sums from Table 7.8 have once again been reduced to a common timescale by calculating the NPV for each item; this time a 3% discount rate has been applied. In this example, you will notice the softwood windows are not redecorated in year 15, as they will be replaced at this stage.

**Table 7.8** Life cycle data for softwood and hardwood windows.

|  | Softwood | Hardwood |
| --- | --- | --- |
| Initial cost of 10NR windows | £3,500 | £6,200 |
| Cleaning cost (per annum) | £120 | £120 |
| Renewal (R) | Every 15 years | Every 30 years |
| Maintenance (M) redecoration | Every 5 years | Every 5 years |
| Redecoration costs | £250.00 | £150.00 |

**Table 7.9** Life cycle cost analysis for softwood and hardwood windows.

| | Cost Heading | r% | n (years) | | Hardwood | Softwood |
|---|---|---|---|---|---|---|
| | **Initial Cost** | | | | £6,200 | £3,500 |
| | **Annual Cleaning** | 3% | 20 | 14.88 | £1,786 | £1,786 |
| HW | £120 x PV of £1 PA | 3% | 20 | 14.88 | | |
| SW | £120 x PV of £1 PA | | | | | |
| | **Redecoration** | | | | | |
| HW | £150 x PV of £1 | | | | | |
| | Year 5 | 3% | 5 | 0.863 | | |
| | Year 10 | 3% | 10 | 0.744 | | |
| | Year 15 | 3% | 15 | 0.642 | | |
| | Year 20 | 3% | 20 | 0.554 | | |
| | £14,000 @ | | Σ | 2.803 | £421 | |
| SW | £250 x PV of £1 | | | | | |
| | Year 5 | 3% | 5 | 0.863 | | |
| | Year 10 | 3% | 10 | 0.744 | | |
| | Year 20 | 3% | 20 | 0.554 | | |
| | £10,000 @ | | Σ | 2.161 | | £541 |
| | **Replacement Windows** | | | | | |
| SW | £3,650 x PV of £1 | | | | | |
| | Year 15 | 3% | 15 | 0.642 | | |
| | £3,650 @ | | Σ | 0.642 | £2344 | |
| | | | | | £8,407 | £8,171 |

## 7.7 Summary

The major focus of the chapter concentrated on the need for the surveyor to look further into a building's life cycle and to predict the costs of maintenance, occupancy and replacement of elements, sub-elements or components. The importance of TOTEX, CAPEX and OPEX were introduced, and the way many organisations manage their asset budgets by having a capital and revenue account has been critiqued, with the argument offered that only by adopting a TOTEX approach and commissioning life cycle cost analysis at the outset of the project would the client achieve maximum efficiency and best overall value from their investment in the building. Finally, the chapter introduced the NRM and the life cycle cost plan, although due to its scale, it was not possible to provide an example life cycle cost plan in this chapter.

The benefits and challenges of life cycle costing have been explored and analysed. In reading this chapter through, the arguments of Mansfield (2009) must be considered. Mansfield is of the view that economic models are as yet unable to assess exactly the time at which a particular building, or group of buildings would be ready for refurbishment. The argument centres on the assertion that whilst mathematical models can assess average life expectances and thus model depreciation, it is impossible to factor in variables such as obsolescence. Whilst this is arguably the case, and the subjectivity of the LCC model is often challenged and seen as the biggest barrier to adoption, it must also be remembered that the use of LCC provides the built asset manager with decision data on which the benefits or drawbacks of increasing capital expenditure to reduce overall life cost can be considered and the optimum trade off determined based on a TOTEX perspective, thus enhancing the value achieved within the project. The lessons learnt from the effective use of LCC in the PFI/PPP sector are important, as they suggest whilst the accuracy issues cannot be ignored, LCC models do provide a basis on which maintenance decisions can be made and resource requirements modelled, with usage in this marketplace leading to situations where the provision of more durable components has been made in designs to ensure they are better aligned with the life expectances of the buildings into which they are incorporated. However, in concluding the chapter, it can be argued that LCC has come along way and is now an important yardstick for value, but there is more work to do to overcome the deeply ingrained business culture that compartmentalises capital and maintenance funding on traditional projects.

# References

Ashworth, A. and Perera, S. (2015). *Cost Studies of Buildings*, 6th ed. Oxon: Routledge.

Building Cost Information Service (BCIS). (2006). *BMI Life Expectancy of Building Components: A Practical Guide to Surveyors' Experiences of Buildings in Use*. London: Building Cost Information Service.

Building Cost Information Service (BCIS). (2011). *BMI Special Report – Occupancy Costs of Specialist School Blocks*. London: Building Cost Information Service.

Constructing Excellence. (2004). Whole life costing. https://constructingexcellence.org.uk/wp-content/uploads/2015/03/wholelife.pdf (accessed 29 May 2021).

Edwards, S., Bartlet, E., and Dickie, I. (2000). *Whole Life Costing and Life Cycle Assessment for Sustainable Building Design*. BRE Digest 452. Watford: BRE Press.

Flanagan, R. (1989). *Life Cycle Costing: Theory and Practice*. Oxford: BSP.

Giuseppe, E.D., Massi, A., and Orazio, M.D. (2017). Impacts of uncertainties in life cycle cost analysis of buildings energy efficiency measures: Application to a case study. *Energy Procedia* 111: 442–451.

Hackett, J. (2020). Final report of the expert group on structure of guidance to the building regulations. https://assets.publishing.service.gov.uk/government/uploads/system/uploads/attachment_data/file/877525/Final_report_of_the_Expert_Group_on_Structure_of_Guiance_to_the_Building_Regulations.pdf (accessed 29 May 2021).

Higham, A., Bridge, C., and Farrell, P. (2017). *Project Finance for Construction*. Abingdon: Routledge.

Higham, A.P., Fortune, C., and James, H. (2015). Life Cycle Costing: Evaluating its use in UK practice. *Structural Survey* 33 (1): 73–87.

HM Government. (2013). Construction 2025: Industrial strategy for construction – Government and industry in partnership. https://assets.publishing.service.gov.uk/government/uploads/system/uploads/attachment_data/file/210099/bis-13-955-construction-2025-industrial-strategy.pdf (accessed 29 May 2021).

HM Government. (2018). Construction sector deal. https://assets.publishing.service.gov.uk/government/uploads/system/uploads/attachment_data/file/731871/construction-sector-deal-print-single.pdf (accessed 29 May 2021).

HM Treasury. (2020). The Green Book (2020). https://www.gov.uk/government/publications/the-green-book-appraisal-and-evaluation-in-central-governent/the-green-book-2020 (accessed 29 May 2021).

Hume, N., Terazono, E., and Raval, A. (2021). Record iron ore price signals raw material boom as economies surge. *Financial Times*, Tuesday 11 May 2021.

Ive, G. (2006). Re-examining the costs and value ratios of owning and occupying buildings. *Building Research and Information* 34 (3): 230–245.

Mansfield, J.R. (2009). Sustainable Refurbishment: The potential of the legacy stock in the UK commercial real estate sector. *Structural Survey*, 27(4): 274–286.

Miller, I. (2014). Life Cycle Costing – Art or Science? Maintenance and Engineering 1st March. https://www.maintenanceandengineering.com/2014/03/01/life-cycle-costing-art-or-science (accessed 29 May 2021).

Mydin, M.A.O., Ramli, M., and Awang, H. (2012). Factors of deterioration in buildings and the principles of repair. *Eftimie Murgy ResltA* 19 (1): 345–352.

NHS Estates. (2001). Sustainable development in the NHS. https://assets.publishing.service.gov.uk/government/uploads/system/uploads/attachment_data/file/147978/Sustainable_Development_in_the_NHS.pdf (accessed 29 May 2021).

Oduyemi, O.I. (2015). Life Cycle Costing methodology for sustainable commercial office buildings. PhD https://core.ac.uk/download/pdf/46171007.pdf. (accessed 6 September 2021)

Pinder, J. and Wilkinson, S.J. (2001). Measuring the obsolescence of office property through user-based appraisal of building quality. CIB World Building Congress, April 2001, Wellington, New Zealand.

RICS (2011). Discounted cash flow for commercial property investments. https://www.rics.org/globalassets/rics-website/media/upholding-professional-standards/sector-standards/valuation/discounted-cash-flow-for-commercial-property-investments-1st-edition-rics.pdf (accessed 29 May 2021).

RICS (2014). NRM3 RICS New rules of measurement: Order of cost estimating and cost planning for building maintenance works. https://www.rics.org/globalassets/rics-website/media/upholding-professional-standards/sector-standards/construction/nrm_3_building_maintenance_works_1st_edition_pgguidance_2013.pdf (accessed 29 May 2021).

RICS (2016) Life cycle costing. https://www.rics.org/globalassets/rics-website/media/upholding-professional-standards/sector-standards/construction/black-book/life-cycle-costing-1st-edition-rics.pdf (accessed 5 September 2021)

RICS (2018). Depreciated replacement cost method of valuation for financial reporting. https://www.rics.org/globalassets/rics-website/media/upholding-professional-standards/sector-standards/valuation/drc-method-of-valuation-for-financial-reporting-1st-edition-rics.pdf (accessed 29 May 2021).

RICS (2020). Professional statement: Cost prediction. https://www.rics.org/globalassets/rics-website/media/upholding-professional-standards/sector-standards/construction/cost-prediction_ps_1st-edition-19-nov-2020.pdf (accessed 29 May 2021).

Swaffield, L.M. and McDonald, A.M. (2008). The contractor's use of life cycle costing on PFI projects. *Engineering, Construction and Architectural Management* 15 (2): 132–148.

# 8

# Sustainable Maintenance Management

## 8.1 Introduction

This chapter introduces the idea and concept of sustainable maintenance management and argues for a change in construction industry practices that embrace the role and importance of the building maintenance professional during the construction stage of works. Having their knowledge and experience during the capital expenditure stage will ultimately lead to a more sustainable and efficient operational stage of a building's life. The circular economy is also introduced and how it is challenging and changing the historically linear thinking of the built environment when it comes to the sourcing of materials. Concepts such as 'carbon neutral' and 'retrofitting' are outlined, and their importance to the role of the building maintenance professional discussed. Reporting frameworks such as the Building Research Establishment Environmental Assessment Method (BREEAM) are also highlighted and how they can be used for the maintenance of buildings explained. The corporate social responsibility (CSR) of building maintenance companies is then discussed as well as the principles of the concept generally and how it can guide the sustainable and socially responsible behaviours of organisations. Finally, the Sustainable Development Goals (SDG) are introduced as a method of bringing together the actions of nations, organisations, and individuals to help navigate the different sustainable options that exist and attempt to focus behaviours for maximum positive impact.

## 8.2 Sustainable Maintenance Management

Sustainability can mean many things to many different people. We all could describe what it means to us and outline our own interpretation, and our interpretations would all broadly align. There will be nuanced differences between different stakeholders' understandings, but not enough to hinder progress for the sustainability agenda. Broadly, and for the purposes of this chapter, sustainability can be described as the ability to meets today's need without compromising the ability to meet future needs. This can be applied to groups of people operating at the same time, and so it is therefore unsustainable for one group of people within society to take actions that will reduce the ability of others in

*Introduction to Built Asset Management*, First Edition. Dr Anthony Higham, Dr Jason Challender, and Dr Greg Watts.
© 2022 John Wiley & Sons Ltd. Published 2022 by John Wiley & Sons Ltd.

society to meet their own needs. This can also apply across generations, and so it will be unsustainable for groups to take actions today that will then prevent future generations from meeting their own needs.

'Maintenance management' is the term given to ensuring that all plant and equipment is regularly inspected, repaired and, where necessary, components replaced to ensure optimum working conditions are preserved and the plant and equipment can continue their operations.

Therefore, the combination of these concepts gives us sustainable maintenance management, in which the management of the maintenance process is carried out in a sustainable manner that focuses on ensuring the operations associated with the maintenance works are not completed in a way that compromises the ability of others to meet their own needs. For building maintenance professionals this translates to the need for a proactive and consistent policy of checking and undertaking maintenance works.

This proactive policy, however, should not start when a building is completed and handed over to a client who will then employ a building maintenance company. This is the 'traditional' approach adopted, but straight away will serve as a disadvantage to building maintenance professionals who will then be responsible for maintaining a building where they have had no prior input into its construction. There needs to be a closer relationship between the professionals involved at pre-construction, construction, and the occupancy stages of a building's life cycle. Whilst the client is the one constant (usually) at all these stages, they are therefore best placed to ensure that professionals who will be involved in the operation stage of a building are involved at all previous stages, as the operation of a building will last much longer than its pre-construction and construction stages. This can be problematic however, as often the client will dispose of the building once complete through a sale, and therefore a new owner will be in place that has had little to no oversight of the pre-construction and construction stages. Nevertheless, as a building's operation last much longer than its construction, it is wise to have building maintenance input into building construction.

Due to the disparate and disjointed nature of a largely competitively driven and contractually minded construction industry, the involvement of building maintenance professionals at earlier stages will not result in a higher premium for the sale of a completed building, and therefore rarely happens. The time has come for such behaviours to change and construction professionals to understand and appreciate the benefits the involvement of building maintenance professionals can have to the buildings end user and ultimate owner. To ensure sustainable maintenance management can be effectively carried out the expertise and understanding of those tasked with maintaining a building can help plant and equipment be procured that is the most effective and efficient over the long term. For example, if a certain make of heating system will require a lower amount of repairs than a similar priced alternative, or if one type of door entrance system uses more readily available spare parts than another.

Utilising this knowledge, the operational costs (OPEX) can be proactively reduced, as can any down time associated with repairs and maintenance works. The current construction industry focus is often purely on the capital costs (CAPEX). Therefore, to truly embrace sustainable maintenance management, the construction industry needs to change its behaviours and adopt a more long-term collaborative approach that considers the operational impacts construction stage decisions can have.

**Figure 8.1**  Linear economy.

The principles of sustainable maintenance management can still be adopted by building maintenance professionals, even if the first involvement they have with a building is when their contract starts once the building is operational. For example, ensuring plant and equipment is regularly inspected and maintained and issues are identified before they become problems and where problems do arise, they are identified and rectified as early as possible. Ensuring the full life cycle costs of replacement parts and materials are also considered, with parts responsibly sourced, are also sustainable maintenance management practices. Adoption of these will extend the life of a building's plant and equipment, and ultimately extend the productive life of a building itself.

## 8.3  Circular Economy

The circular economy is essentially an idea that focuses on the reuse and recycling of resources. Such an approach was developed in contrast to the traditional and historic view that resources are extracted, and then in a linear model, processed and used before being discarded, as can be seen in Figure 8.1.

The linear economy approach was developed at a time of substantial growth and industrial development, when little was known about the true finite nature of resources and raw materials and the full extent of environmental and social damage was not yet understood. In the linear economy model the focus is always on short-term economic growth. Little to no focus was afforded to the discard stage, not in the long term anyway. As if any real consideration was afforded, then it would have become apparent the constant generation of waste was unsustainable. This is a somewhat extreme and negative consideration, as in practice, research has been undertaken regarding the discard phase and questions have been asked regarding the sustainability of constant waste generation. It just took a little time for answers to these questions to be fully understood and for this knowledge to spread throughout society. Indeed, it was only in 2015 that a charge was introduced for single use carrier bags, and only in October 2020 that single use plastics were banned in the UK. This pattern will continue with more initiatives introduced going forward. Such initiatives, however, are part of what can be described the circular economy.

Recycling and the reuse of materials are key parts of the circular economy, but it is about a paradigm shift away from the linear economy model and ultimately ensuring that the discarding of resources is minimised, and wherever possible the discard phase is linked to, or replaces, the extract phase, as can be seen in Figure 8.2.

Whilst extraction of raw materials will always be required, the aim of the circular economy is to reuse all resources where possible, and then recycle the resources once they have been reused to the limit of their lifespan. The circular economy is of the upmost importance to the building maintenance sector. In order to ensure sustainability of any refurbishments and ongoing maintenance it is key that the circular economy is

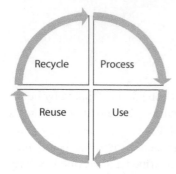

**Figure 8.2**   Circular economy.

understood, and resources are reused wherever possible. When resources do have to be purchased, ensuring resources are made either entirely or partially from recycled and reused materials is key to helping any sustainable credentials and targets are achieved. For some resources reuse may be difficult if not impossible, and it can be costly (prohibitively so) for recycling to take place. In cases like this, alternative sourcing should be attempted so extraction of the raw materials involved in the resource can be minimised. Where such instances cannot be avoided, a form of 'offsetting' may be explored. For example, if raw timber is required, procuring this from FSC (Forest Stewardship Council) sustainable sources and planting additional trees to cover those trees used in the maintenance works will be a suitable sustainable solution.

Whilst there has been some movement towards a circular economy, a much more focussed and widely adopted approach will be required across the building maintenance sector to ensure sustainability targets are achieved. Taking a circular economy approach needs to be the standard behaviour of all building maintenance companies and the owner of all build environment assets. It can therefore be considered one of the responsibilities of a building maintenance professional to be informed of the circular economy benefits and how the principles can be embedded in their current project(s). This knowledge then needs to be communicated to the building owner, both in submitted tenders, and during any building maintenance contracts currently underway. If raw resource use can be minimised and reuse and recycling rates improved, then they should be implemented as a priority and conducted as part of a business-as-usual approach. Table 8.1 outlines some practices that can be undertaken by building maintenance professionals to help embed principles of the circular economy in their practices.

Where full circular economy principles have been adopted at a building's design, construction, and maintenance, then a building may be considered as carbon neutral.

## 8.4   Carbon Neutrality

'Carbon neutrality' refers to a calculation between the carbon that is emitted and the carbon that is removed. This removal can be accomplished via a combination of reduction and offsetting. Once the result of carbon reduction and offsetting is equal to the amount

**Table 8.1** Circular economy principles for building maintenance professionals.

| | |
|---|---|
| Water use monitoring and reduction | Having accurate and reliable water monitoring systems in place is key to understanding a building's water use. As well as maximising grey water capture and reuse systems, reducing initial water consumption is key for any building. This can only be achieved by a building maintenance professional being fully aware of where, when, and how a building consumes water. When the full facts are known and understood regarding a building's water use, plans can then be put in place to effectively minimise consumption. |
| Waste reuse policies and promotional materials | Having reuse policies in place are an effective start, but these then need to be adopted and undertaken by a building's users. These could be children at a school a building maintenance professional is managing, or residents at a block of flats. Promoting the policies to enable and encourage widespread adoption is then the next stage to facilitate their success. This could include any food composting practices in shared gardens and access to an onsite furniture exchange whereby residents can donate and adopt furniture rather than disposing of items. |
| Waste segregation and recycling policies | At any building that has multiple users, getting them all to follow the same set of policies can be difficult. However, having clearly marked separate bins, with labels or a colour coded system in place can make it easy for users to recycle. Having extensive recycling relationships in place with external providers can also educate building users to just how many different 'waste' products they previously generated can be recycled with the multitude of bins you will have available. |
| Switching to alternative materials | Where materials need to be consumed and disposed of, efforts should always be made to ensure sustainable products are used. This can range from a toilet roll that derives from sustainable timber forests to cleaning and food products that do not have ingredients which originate from endangered habitats and ecosystems. |
| Utilise smart and innovative technology | This can range from heat recovery systems to smart meters monitoring energy use and the efficiency of a buildings equipment. By adopting and utilising the latest innovative technology available, buildings can ensure they are at the forefront of the sustainable revolution and are as efficient as they can be. |

of carbon emitted, 'net zero' has been achieved. This is the target many individuals, companies, industries and even countries are trying to achieve. If the amount of carbon removed exceeds the amount of carbon emitted, then it is described as carbon positive. There are many international targets set for countries, and national and regional targets set for both public and private sector companies to achieve net zero. The building maintenance sector is no different.

There are numerous activities that building maintenance companies and professionals can undertake to help achieve their carbon targets. These include ensuring they are aware of the current performance of a building and are measuring its current carbon performance accurately. It is only by having clear, transparent accurate and robust measures and metrics in place that a building's performance can be truly understood. Once a clear measure of current performance is achieved, the next stage is to understand a building's potential. This should then be mapped so it is clear the targets that need to be achieved. A series of targets should then be set to enable the building's carbon performance to be increased from its current benchmark level to reach its full potential.

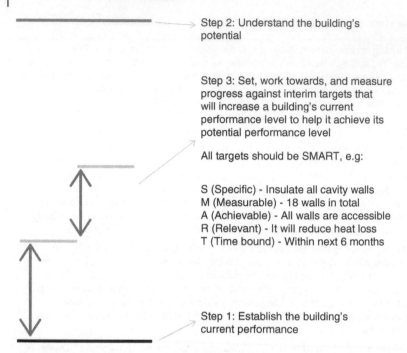

Step 2: Understand the building's potential

Step 3: Set, work towards, and measure progress against interim targets that will increase a building's current performance level to help it achieve its potential performance level

All targets should be SMART, e.g:

S (Specific) - Insulate all cavity walls
M (Measurable) - 18 walls in total
A (Achievable) - All walls are accessible
R (Relevant) - It will reduce heat loss
T (Time bound) - Within next 6 months

Step 1: Establish the building's current performance

**Figure 8.3** Steps to achieving carbon neutrality.

One method that can be employed when setting such targets is to ensure that all targets are SMART. This is an acronym for Specific, Measurable, Achievable, Relevant and Timebound. It is important that the targets are all SMART, as this will help improve their chances of being successfully achieved, as illustrated in Figure 8.3.

In addition to the carbon reduction techniques building maintenance professionals can adopt for their buildings, one carbon offsetting technique that is sometimes used is the method of one company 'purchasing' carbon 'credits' from another company. This is where company A exceeds their carbon target, and company B does not exceed their carbon targets. The 'room' between company B's actual carbon emissions and their targeted allowed omissions is then sold to Company A. Effectively, Company B reduces their targets so that Company A can increase theirs. This results in additional income for Company B and allows Company A to say they are meeting current carbon targets. This method does have its critics as arguments are put forward that Company A is effectively buying its way out of its obligations; however, if the average effect is that carbon emissions are reduced across both companies then others argue it is a beneficial arrangement, as it helps contribute to the reduction in national levels of carbon in the atmosphere.

The negative impacts of increased levels of atmospheric carbon are well documented, from increased pollution and pollution-related diseases to negative impacts on ecosystems. The global warming of the atmosphere to historic levels from which the earth will not recover is now a well-known and rapidly occurring phenomena that will take positive action, both reactive and proactive from multiple stakeholders simultaneously to address. Building maintenance professionals are equally responsible for the fight against climate change through the carbon neutral agenda. One of the main methods by which building maintenance professionals can help achieve the carbon neutral targets is by undertaking an expansive programme of retrofitting to the existing building stock.

## 8.5　Retrofitting

Retrofitting is the process whereby existing buildings have new technology installed during their operation phase. In essence, in the built environment it is the fitting of non-original components to enhance or repair an existing asset. Retrofitting has been described as

> providing something with a component or feature not fitted during manufacture or adding something that it did not have when first constructed. The term has been used in the built environment to describe substantial physical changes at building level and has often been used interchangeably with terms such as 'refurbishment', 'conversion' or 'refit'. But at an urban or city scale retrofit means something much larger and more comprehensive, more integrated; underpinned by sustainable financing and with a clearly defined set of goals and metrics. The Retrofit 2050 project therefore defines sustainable urban retrofitting as the directed alteration of the fabric, form or systems that comprise the built environment to improve energy, water and waste efficiencies.
>
> *(Eames et al. 2014)*

This can be for a multitude of reasons, such as to bring buildings in line with the latest legislative requirements and stakeholder expectations. It could also be to maximise the return on investment that new technology often affords. For example, if a hotel building was completed 20 years ago, the heating and air conditioning units in each room will have been recent releases at the time of installation, but will now be outdated technology. The latest technology will offer much greater performance benefits that will result in cost savings when it comes to the energy costs associated with each room when occupied. Therefore, although a high initial upfront cost may be incurred for the retrofitting of heating and air conditioning units, the increase in performance and decrease in running costs could provide a return on investment in just a few years.

Repair and maintenance work carried out as and when required, for example, if the hotel above had a damaged door to one of the rooms that needed to be replaced, this could not be described as retrofitting. Nor would the replacement of damaged room furniture or damaged light switches. The replacement of all light bulbs however, with energy efficient smart lighting would fall into the category of retrofitting, as would the replacement of traditional light switches with lighting sensors. This distinction needs to be understood so that works can be accurately categorised. Retrofitting with new energy efficient technology may be subject to government grants and initiatives and so being aware what falls within such categories could be key to maximising any applicable financial support available. Whilst this is often primarily aimed at domestic residential properties, commercial properties do sometimes qualify for such grants and initiatives.

In addition to potentially securing funding to support retrofitting opportunities, there are numerous other benefits a building and its stakeholders will experience by retrofitting.

This includes prolonging the useful life of a building. If key elements of plant and equipment have been maintained well and retrofitted with the latest technology as and when it became available, then the average age of building components will be much newer than the age of the building and so the condition and quality of the building should be sufficient to continue operating beyond its original anticipated end-of-use date. Bringing the building's plant and equipment up to date also allows for the availability of spare parts as and when the plant will need replacing. For example, a central heating system installed 20 years ago will consist of 20-year-old parts. Should the manufacturer have gone out of business or the model of heating system be discontinued, it may prove prohibitively expensive, even impossible, to secure replacement parts for any repair works required to be undertaken. In such instances, having frequent retrofitting of newer technology will result in the natural replacement of older parts with newer, more freely available ones.

In March 2014, 'Retrofit 2050: Critical Challenges for Urban Transitions' was released, which identified key success factors that need to be in place to ensure sustainability can be achieved across UK cities (Eames et al. 2014). Table 8.2 describes the authors' interpretation of the eight key success factors that are described as essential to achieving this goal:

Ultimately, retrofitting with newer technology should improve the welfare, safety, and experience of a building by its users. It can lower the emissions generated and improve

**Table 8.2** Retrofit 2050 eight Key Success Factors.

| | |
|---|---|
| 1. An inclusive urban retrofit agenda | The differing motivating factors of retrofitting need to be understood, these can include economic and social motivations. For example, to encourage private investment in a region that is seen as taking the retrofit agenda seriously, or as part of a contribution of wider social responsibility to achieving the sustainability agenda. |
| 2. Compelling retrofit city visions | A clear sustainable vision is required for each city's future to manage the competing pressures retrofitting on a large scale will bring. A recognition of the individual demands each city will face as well as bespoke dynamics of their demographics and diversity will need to be considered. |
| 3. Improved modelling and decision support tools | Due to the complexity of the built environment, a current barrier to rolling out retrofitting on a large scale and monitoring the success is the lack of modelling and decision supporting tools. Whilst some progress has been made with the modelling of buildings, more development is needed to accurately reflect the challenges retrofitting poses. |
| 4. Institutional capacity, planning and governance | There is often a disconnect between short-term targets and long-term planning. Integrated planning across disciplines and amongst stakeholders is required with a greater focus on sustainability at every stage and an appreciation of whole life costs, not just capital costs. |
| 5. Access to 'green' finance | Financial incentives and instruments need to be freely provided to encourage retrofit take up and achieve the sustainability goals set. The disconnects between building owners and operators needs to be bridged and a greater focus and appreciation of the importance of retrofitting needs to be gained. |
| 6. Effective partnerships | Private and public sector partnerships will help achieve the goals set and so it is important such avenues are pursued both in the creating and delivery of strategies and capturing of data to better inform future actions and investment decisions. |

*(Continued)*

**Table 8.2** *(Continued)*

| | |
|---|---|
| 7. Long-term sharing of risks and benefits | Current business models favour short-term investment decisions and so the realisation of any longer benefits may not be attractive to some parties under current approaches. A shared long-term approach to risk management and the realisation of associated benefits is required for the sustainable agenda to be realised. |
| 8. A whole systems perspective | Many of the benefits realised to date have been gained from strategies impacting individual residential and commercial properties. To ensure greater benefits can be achieved in the medium and long term an approach is required that addresses the whole system of retrofitting and the built environment. |

the safety features available, it can also help existing buildings serve new purposes. If an old warehouse or commercial unit is derelict and has no immediate potential to be occupied, then retrofitting suitable technology can turn the premises into a habitable place which could then result in the buildings being effectively repurposed as residential properties. Table 8.3 expands on some of the opportunities and benefits retrofitting provides for building maintenance professionals.

**Table 8.3** Opportunities and benefits of retrofitting.

| | |
|---|---|
| Security | CCTV technology has improved dramatically over recent years, and so outdated, oversized and hardwired CCTV that may have been installed when a building was first complete can now be updated. CCTV can now include wireless indiscrete wide-angled devices that offer greater clarify, have in built battery backups and can save direct to cloud storage, negating the need for an internal building location to store. Retrofitting with such a system would ensure parts are more easily replaced, the system is more secure and the security coverage can be increased. |
| Lighting | Traditional building lighting systems used to operate room lighting can be retrofitted so that switches are removed and replaced with sensors. This has been undertaken in many schools and portable construction cabins. Such retrofitting can reduce the costs of replacement and repair due to damage of features such as light switches, and also enhance safety and security as no lights can be turned off when they should not be. It can also reduce energy bills, as lights will not be left on accidentally, and ensure buildings are more sustainable and energy efficient. |
| Insulation | Enhancing a building's thermal properties is one of the main ways retrofitting to current buildings takes place. This can include the addition of external insulation and render, or can be through methods such as spraying insulating foam into an existing cavity wall. The benefits of this include lower energy costs and energy wastage, as well more ambient indoor temperatures and environments. |
| Windows | Windows can be replaced with double- and even triple-glazed units to enhance thermal properties and make the building more efficient maintaining temperatures and air quality. This can also reduce draughts and solar glare and so can have positive impacts on the well-being of a building's users, too. |
| Renewable energy sources | One of the more visible and easily identifiable elements of retrofitting new technology to existing buildings can be the use of solar panels or wind turbines. These are affixed externally to buildings and can split opinion on the attractiveness of their appearance, yet the benefits they offer can be quite immense. From a reduction in energy costs and an increase in the total amount of renewable sources buildings use. Ultimately this will reduce the burden on finite fuel sources and the environmental impacts associated with the extraction of these resources. |

When retrofitting takes places, it is often one item at a time and helps a building slowly evolve and improve over its lifetime. This would perhaps fall under the category of kaizen, which is to continuously improve and develop. However, a single major retrofitting 'event' could also occur whereby a building is effectively closed to its normal use whilst retrofitting occurs. This would be where a major process of retrofitting is being undertaken that makes the building temporarily inoperable or poses significant risk to stakeholders. Examples could include the transformation of a building from one purpose to another (such as commercial unit to residential). When retrofitting does occur, there are standards and initiatives that may be adopted by the building maintenance professionals involved, or enforced by clients and building owners. Such standards and initiatives serve to improve the performance of the building, both during the refurbishment phase, and in its post-refurbishment operation. BREEAM is one standard that is often adopted to assess the sustainable and environmental credentials of a building's refurbishment and operation.

## 8.6 BREEAM

In 1990 the Building Research Establishment (BRE) introduced the Building Research Establishment Environmental Assessment Method (BREEAM). BREEAM is ultimately a method of assessing the environmental and sustainable performance of both new build and refurbishment projects. It assesses the different stages of a project against a range of categories including Energy, Health and Well-being, Land Use and Ecology, Management, Materials, Pollution, Transport, Water, and Waste. An independent assessor will assess each scheme against the criteria in each category and the project will be awarded a rating of either Unclassified (having scored less than 30%), Pass (30–45%), Good (45–55%), Very Good (55–70%), Excellent (70–85%), and Outstanding (85%+).

BREEAM has been an excellent driving force in increasing the sustainable credentials of both new build and refurbishment projects, and arguably either directly or indirectly impacts all construction projects and properties. If a new build or renovation is registered with BREEAM then the impact is direct and obvious, as the project will be judged against the criteria set by BREEAM. However, even if the project is not BREEAM registered, BREEAM has served to increase the general expectation of all buildings' sustainability credentials and so as a building maintenance manager you will have increased sustainability focused responsibilities.

The ultimate aims of BREEAM are to mitigate the impact a building's life cycle has upon the environment, to encourage and enable the recognition of the environmental benefits of buildings and to stimulate the market demands for a sustainable built environment. The assessments undertaken span numerous criteria, but in the authors' view, examples applicable to the building maintenance sector are outlined in Table 8.4.

Achieving a high BREEAM score is a successful accomplishment that the entire team should be proud of. When such accomplishments are achieved, the company's involved will often publicise the BREEAM participation and score and use the project as a good news story and evidence of their abilities and performance in future marketing and

**Table 8.4** BREEAM criteria and potential Building Maintenance impacts.

| | |
|---|---|
| Health and Well-Being – Indoor Air Quality | A building maintenance manager will need to ensure that all windows are functional and can be securely opened to allow adequate ventilation. Air conditioning and temperature units will need to be fully functional and fit for purpose with the wider geographical factors considered, such as the angle of the sun on a building façade and proximity to busy roads or industrial areas, so that internal air quality can remain at optimum levels. |
| Health and Well-being – Safety and Security | When considering the occupants and users of any building, their safety and security must be considered at all times. Whether initiatives such as Secured By Design are adopted to enhance security and/or security measures are implemented post-completion in relation to additional concerns, the welfare of stakeholders must be a primary function in the operation of a building. This also extends to during any repair works that are required, with a need for a safe approach to be adopted by all professionals to minimise disturbance and risk of harm to all stakeholders. |
| Energy – Reduction of energy use and carbon emissions | Whilst many of the sustainable features will have been installed during the construction of the building, those tasked with building maintenance will have to ensure all sustainability features remain fully functional during operation. They will also need to ensure performance of any innovative technologies is monitored and any outputs lower than planned need to be reported and addressed in a timely manner. Retrospective additions to existing buildings also form a significant part of the building maintenance teams role to ensure a building's performance improves in line with changing stakeholder expectations and utilises the latest technology where available. It is of paramount importance an awareness of the latest technological introductions and developments in practice is maintained as well as the latest legal requirements for the performance of buildings and any appropriate initiates to improve performance. |
| Transport Cyclist facilities | It is important that cycle facilities, where installed, are maintained correctly and adequate security is in place. Where these are not installed when the building is constructed, they can be reactively installed. The building maintenance team therefore need to ensure the location is suitable and the addition complements the building and its requirements and inhabitants/occupiers. |
| Water – Water consumption/ monitoring | The amount of water consumed by buildings is under increasing scrutiny and so building maintenance teams need to ensure there are no leaks and any problems that do arise are able to be reported quickly and addressed safely. The consumption of water needs to be recorded and monitored on an ongoing basis to ensure it is being used as efficiently as possible, and all activities that require water use, such as cleaning of windows, etc., are not carried out in a wasteful manner. |
| Materials – Responsible sourcing of materials | Whenever any repair or refurbishment work is taking place it is of the upmost importance that all materials procured are done so from responsible sources. For example, FSC timber evidences it is sourced from responsible sources and so ensuring all timber has FSC certification is a must, as it is ensuring the ethical behaviours and legal compliance of all supply chain partners. This extends to labour practices and the need to ensure standards are evidenced and maintained. |
| Pollution – Flood risk management and reducing surface water run off | The risk of flooding occurring to all buildings appears to be increasing. Extreme weather events are occurring with increased frequency and so building maintenance teams need to be aware of, and prepare for, dealing with such instances. Events shouldn't be waited for and then dealt with retrospectively. Whilst this always forms a part of reflection and lessons learnt for risk management purposes, a proactive approach should be taken. The risk of flooding should be assessed and appropriate action taken to reduce the likelihood of extreme weather events having an overly adverse negative impact. |

tendering opportunities. This will then form part of a company's broader corporate social responsibility (CSR) strategy and help the companies involved be recognised for the positive sustainability and environmental impacts they are responsible for.

## 8.7 Corporate Social Responsibility

Corporate social responsibility (CSR) is a term that is used to describe the wider responsibility of an organisation. It can encompass both behaving in a responsible way for day-to-day operations, and going above and beyond what is necessary for the basic operation of an organisation. For example, a building maintenance company may decide to hire certain disadvantaged groups and get all their materials from responsible and locally sourced suppliers. This would be classed as behaving in a responsible manner. If the same company were to then donate a certain percentage of their profits to charity, this could be classed as going above and beyond what is generally required for them to complete their core business responsibilities and obligations.

As there is no widely agreed definition of what CSR relates to, the range of CSR behaviours a company can adopt can and will vary. A building maintenance company can decide to adopt a firm and somewhat rigid position on CSR, with their own uniquely created definition that makes sense to their staff and operations and can be used to guide actions and strategies towards a single purposeful aim. For example, a building maintenance company may decide that to them, CSR is supporting a charity aimed at tackling homelessness that is local to their head office. So, therefore, the building maintenance company may then raise funds for the charity, donate a percentage of their profits to the charity and offer training opportunities, as well as providing work experience and permanent positions to those classed as homeless whom the charity seeks to support. The benefits of this will undoubtedly be a greater positive impact felt by the charity. The sense of improvement and benefit will also be felt amongst the work force, who may be invigorated by a sense of purpose and encouragement at seeing this positive difference and be inspired to continue. The drawbacks however, are that other charities may be in equal need of funds and support but having not been selected by the building maintenance company to be a charity partner may not be able to continue to provide the services they are depended upon for. Such an action may also actually reflect badly upon the building maintenance company itself. For example, if the company tendered for a contract and was comparable across several criteria to their main competitor, but the client had a national profile. If both building maintenance contractors had a CSR policy, but the first was local to one area (the homelessness charity) and the second company committed the same resources to their CSR as the first, but had a national CSR agenda, then the latter company could be seen as sharing a CSR policy matching that of the client's interests.

There are some companies that still view CSR as an 'extra'; as something that should be treated as additional to their core business operations and not intrinsically linked to the manner in which the company operates. Such companies have a somewhat outdated view of business, and whilst may still survive if they are in a niche market, or a supplier of such bespoke importance, or operating in some form of oligopoly, ultimately most companies need to adopt and embrace CSR policies to meet contemporary stakeholder demands and compete in an ever-more-crowded market place. Arguably, all CSR behaviour is now an expectation all stakeholders have of modern-day businesses. It is increasingly important to all stakeholders that businesses do 'the right thing' in the way they operate, so much so that a company who does not engage with CSR practices will find they are at a disadvantage when being compared in any procurement comparison against a company that embraces CSR.

Therefore, all building maintenance companies should have a CSR policy in place, and not be shy to advertise the positive impacts and contributions they make to any immediate recipients and wider society in general. Many companies have regular CSR communications, and when arriving on the landing page of many company websites, you could be forgiven for at first not being entirely sure the company operates in the construction or maintenance world. The first information website visitors often see is the good news CSR policies and stories the company has recently been involved in. This illustrates the importance which some companies place on their CSR and represents the expectation of what some stakeholders perhaps view as the 'standard' of behaviour required.

Whilst definitions of CSR are diverse, there are activities that can be included within the CSR banner, and activities that should not. CSR has three broad categories under which activities can be classed: economic, environmental and social. Whilst actual activities under these headings can be purely philanthropic or conducted as part of day-to-day business transactions, these are the headings under which CSR is usually reported. Table 8.5 provides examples of the initiatives in a building maintenance company.

Arguably, CSR is no longer an optional undertaking from building maintenance companies. Increased expectation from a range of stakeholders is often enough of a factor on its own to motivate companies to engage with CSR – for fear of not maintaining a positive public image. However, there is now an increased focus within procurement, with clients increasingly using CSR factors with greater weighting. Therefore, building maintenance companies arguably need to engage with CSR in order to successfully win work. This is even more of a requirement

**Table 8.5** CSR initiatives of Building Maintenance companies.

| | |
|---|---|
| Economic | This can range from donating profits to charity, encouraging staff to volunteer their time to good causes and procuring materials from sustainable and local suppliers. Numerous case studies and measurement tools exist (although plenty are not as good or accurate as they would lead us to believe). Such tools show that concentrated local spend can have a multiplying effect and each pound spent can be worth more to the local area with repeat spending. |
| Environmental | Activities include reducing pollution emitted from activities undertaken by adopting alternative working practices, and actively serving to protect and enhance biodiversity by planting trees and protecting wildlife. Encouraging staff car sharing and bike to work schemes ensuring sustainable material is sourced are also effective methods to increase a company's environmental credentials and positive impact. |
| Social | In additional to ensuring local and concentrated spending when procuring labour, plant and materials, a company can actively look to recruit staff from underrepresented societal groups. A company could also help clean up local areas, provide work experience opportunities and visit schools and community groups to deliver training and talks. |

when tendering for public sector work since the introduction of the Public Services (Social Value) Act (2012). The Act places a legal obligation upon public sector clients to consider wider criteria than simply time, cost and quality when awarding contracts. This can include the facilities management contracts or repair works on public sector buildings and assets.

Whilst CSR can ultimately be left to each individual company, even if directed by the procurement criteria of clients or expectations of stakeholders, there are still attempts to focus the energies and efforts of companies (as well as governments and other bodies) to achieving those areas deemed as most in need of focus and improvement. Such attempts include the publication of the Sustainable Development Goals.

## 8.8 Sustainable Development Goals

Founded in 1945, the United Nations (UN) is an intergovernmental body that aims to foster and encourage collaboration and cooperation across nations, maintain international peace and security, develop friendly inter-country relationships and act in a manner to harmonise global action. It has been described as the most internationally representative intergovernmental body in the world and in 2015 launched the 2030 Agenda for Sustainable Development.

At the heart of the 2030 Agenda for Sustainable Development are the 17 Sustainable Development Goals (SDGs). Table 8.6 outlines each of the 17 SDGs.

The SDGs are all interlinked, and the success of one is ultimately linked to success of another. The SDGs have been described as an urgent call to action for all nations to tackle some of the greatest challenges the world and its population is facing today. Building Maintenance and Asset Management companies have a great opportunity (and arguably responsibility) to help contribute to the achievement of the SDGs in their organisational and operational behaviours.

For example, internal organisational practices need to address any issues that arise under SDG 5 Gender Equality and ensure there is appropriate representation in the

**Table 8.6** Sustainable Development Goals.

| 1. No Poverty | 2. Zero Hunger | 3. Good Health and Well-Being |
|---|---|---|
| End Poverty in all its forms everywhere | End hunger, achieve food security and improved nutrition and promote sustainable agriculture | Ensure healthy lives and promote well-being for all at all ages |
| 4. Quality Education | 5. Gender Equality | 6. Clean Water and Sanitation |
| Ensure inclusive and equitable education and promote lifelong learning opportunities for all | Achieve gender equality and empower all women and girls | Ensure availability and sustainable management of water and sanitation for all |

*(Continued)*

**Table 8.6** *(Continued)*

| 7. Affordable and Clean Energy | 8. Decent Work and Economic Growth | 9. Industry Innovation and Infrastructure |
|---|---|---|
| Ensure access to affordable, reliable, sustainable, and modern energy for all | Promote sustained, inclusive, and sustainable economic growth, full and productive employment and decent work for all | Build resilient infrastructure, promote inclusive and sustainable industrialisation and foster innovation |
| 10. Reduced Inequalities | 11. Sustainable Cities and Communities | 12. Responsible Consumption and Production |
| Reduce inequality within and among countries | Make cities and human settlements inclusive, safe, resilient, and sustainable | Ensure sustainable consumption and production patterns |
| 13. Climate Action | 14. Life Below Water | 15. Life on Land |
| Take urgent action to combat climate change and its impacts | Conserve and sustainably use the oceans, seas, and marine resources for sustainable development | Protect, restore, and promote sustainable use of terrestrial ecosystems sustainably manage forests, combat desertification, and halt and reverse land degradation and halt biodiversity loss |
| 16. Peace, Justice, and Strong Institutions | 17. Partnerships for the Goals | |
| Promote peaceful and inclusive societies for sustainable development, provide access to justice for all, and build effective, accountable, and inclusive institutions at all levels | Strengthen the means of implementation and revitalise the global partnership for sustainable development | |

company employees and management. This practice should then be encouraged throughout the supply chain with best practice guidance and targets for all supply partners to aim for. Whilst this may be difficult to achieve in some instances, especially if you require works carried out where only a few companies are available and none meet the criteria you have set; nevertheless, it is still a best practice to aim for and should be encouraged, especially in the construction industry where gender inequality has historically been an issue many companies have experienced.

SDG 12 Responsible Consumption and Production is ideally suited to be tackled by the building and asset maintenance sectors as they traditionally use large amounts of raw materials and natural resources in their day-to-day operations. Ensuring timber and bricks are responsibly sourced is an effective start, but also minimising the use of water in cleaning during operations such as window cleaning and ensuring buildings have solar panels or grey water harvesting facilities are effective methods of positively contributing to achieving this SDG.

Whilst progress towards all SDGs can be positively contributed to by the building and asset maintenance sectors, another specific example included here is for progress towards

SDG 15 Life on Land. For example, refurbishing an overground gas pipe that crosses a culvert in the middle of a field will involve gaining access to the remote area with plant, vehicles, and equipment. This will result in disturbance to all local ecosystems in the vicinity and could lead the refurbishment team to encounter endangered or protected species. Such an area may also be overgrown with vegetation and so care must be taken to ensure such vegetation is sustainably cut back. Correct surveys must be conducted to ensure the risk of biodiversity loss is reduced. Whilst much more can obviously be done towards achieving this SDG in countries where desertification is occurring, all construction industry works in all countries can contribute to ensuring SDG 15 is achieved.

## 8.9   Conclusion

The role of the building maintenance professional is challenging and constantly changing with increasing requirements pertaining to the sustainability of buildings. Unfortunately, this role is made somewhat harder with the construction industry's current approach of a capital expenditure (CAPEX) focus at the expense of operational costs (OPEX). This approach manifests itself in building maintenance professionals not becoming involved in a building until the operational stage, and so the expertise, knowledge, experience, and insight they can offer is often lost – ultimately, they are given a building to maintain for which they have had little input into its design and construction decisions. Nevertheless, buildings can still be maintained in a sustainable way, with concepts such as the circular economy and carbon neutrality adopted, and actions such as retrofitting undertaken and embraced. Tools such as BREEAM exist that can help guide the actions of professionals during the building maintenance process, and the adoption of Sustainable Development Goals (SDGs) can provide shared international targets that can encourage collaborative behaviour and offer a guidance framework for companies and buildings looking to adopt targets that aim for a more sustainable future.

## Reference

Eames, M., Dixon, T., Lannon, S., Hunt, M., De Laurentis, C., Marvin, S., Hodson, M., Guthrie, P., and Georgiadou, M. (2014). Retrofit 2050: Critical challenges for urban transitions. https://www.retrofit2050.org.uk/wp-content/uploads/sites/18/2014/04/critical_challenges_briefing-March-2014.pdf

# 9

# Risk Management

## 9.1   Introduction

How building maintenance professionals encounter risk, both contractually, and the physical risks involved in the maintenance of buildings, is explored in this chapter. Initially, this chapter introduces the concept of risk. The traditional view of risk has always been somewhat pessimistic, that risk is a negative phenomenon consisting of only a downside. However, there is upside risk, and risk that provides an opportunity – sometimes this is the same event that provides an upside or downside for a party, or that provides an upside for one party and a downside for another. Whilst there are many methods by which to define risk, ultimately, it is about the probability and impact of an unexpected event. It is only through increasing our understanding of risk against these headings (probability and impact) that the true nature of risk can be explored. Such an exploration can help the correct actions be taken with regards to the retaining or transferring of risks. This is essential in the current climate of the built environment when it comes to the maintenance of buildings and assets. This chapter discusses the tools and techniques that can be adopted by building maintenance professionals to help protect themselves (and the buildings they maintain) against excessive and unnecessary risk exposure.

## 9.2   What Is Risk?

Risk can be described as an uncertain future outcome. A situation in which there is some exposure to an element of chance or danger. Where the end result is unknown, or where one of several end results is certain but there is no way of knowing which. Risk has been defined as 'the likelihood of an event or failure occurring and its consequences or impact' (RICS 2012). Any consideration of risk should include the chance of an event occurring, and the impact and ramifications of that occurrence, should it occur.

Historically, risk has been viewed somewhat pessimistically as a negative. Something to fear and protect against at all costs. Think of the likelihood of a piece of equipment failing in a building; as part of any maintenance schedule this piece of equipment may be

*Introduction to Built Asset Management*, First Edition. Dr Anthony Higham, Dr Jason Challender, and Dr Greg Watts.
© 2022 John Wiley & Sons Ltd. Published 2022 by John Wiley & Sons Ltd.

regularly inspected and maintained (to try and minimise the risk of it failing). The same piece of equipment may also be insured against failure (in an attempt to pass the risk on to someone else). In these circumstances, what is happening is that a third party is taking that risk of repairing or replacing the piece of equipment in return for a fee. You are, effectively, paying someone to take the risk on your behalf. This same approach has been adopted throughout the construction industry, including the maintenance and management sectors. When procuring contractors on a framework agreement of repairs, the contract will (should) clearly state who will be responsible for the risks that exist. Whilst not all risks can or will be identified at the stage the contract is signed, the contract should have clauses in place that deal with how any future risks will be allocated. Whilst this approach has again always taken the view that risks are negative, something to pass on to others, risk can in fact be viewed as a positive phenomenon. This is not to say that all risks can be viewed both positively and negatively; some risks are ultimately negative, and so when transferred, the party responsible for the negative risk should be fully aware, and have been able to price the responsibility of this risk accordingly. However, there does exist positive risk.

Unlike negative risks in which there is a worse than expected (downside) outcome, positive risks are those that can lead to a better than expected outcome (upside). Indeed, this can be categorised as an opportunity. Such a view on risk is rather contemporary and not yet widespread in the construction industry – but this is slowly changing. Parties are starting to see the upside risks and opportunities that exist. There are also those risks that can be both positive and negative. These could be for different parties – a building owner may pass the risk of repairing an item to a maintenance contractor in return for a some of money, and if the item does not need repairing, then that sum of money becomes an opportunity for the maintenance contractor to increase their profit margins. In this situation, both parties are treating the same exchange as the potential to experience both upside and downside risk.

Whichever perspective risk is viewed from, an event will need to lead to a significant consequence in order to be considered a risk. This is where the terms 'probability' and 'impact' start to take on significance.

- Probability – The likelihood an event will occur.
- Impact – The outcome that will be experienced.

If something has such a low impact value, i.e. even if it did occur it would not result in any event that would cause anyone harm or inconvenience, then it is not generally considered a risk. Only events that have the potential to cause an impact of some significance would be discussed in the context of being a risk worth focusing upon. Probability, however, is a different category of risk. There are some arguments of the view that events with such a low probability of occurring are not deemed risks as they are so far remote – and likewise for events that have such an odds-on chance of occurring. It is the view of this book however, that any identified event, regardless of its probability, once identified, needs to be considered a risk. Take for example a pandemic induced lockdown. Who, in 2019, considered that to be a likely event?

## 9.3   The Nature of Risk

Risk, by its very nature, is unexpected. If the impact is known the probability may not be. Similarly, if the probability is known, it is the impact which may be unknown. To repurpose a famous quote by Donald Rumsfeld, risks can be considered as falling into one of the following three categories:

1.  Known knowns – Where the likelihood and impact of an event is known.
2.  Known unknowns – Where either the likelihood or impact of an event is known.
3.  Unknown unknowns – Where both the likelihood and impact of an event is not known.

When it comes to assessing and understandings risks, especially in terms of construction and maintenance risks, the aim is to ensure the nature of all identified risks are understood as best they can be. To try and ensure all risks are known knowns would be the ideal position to be in when it comes to the understanding of risk, as it would allow full preparation to be undertaken on how best to mitigate any negative risks and maximise any positive risks. Figure 9.1 better illustrates the form this type of risk consideration could take.

To better understand the nature of any risks faced, an exercise could be undertaken whereby all risks are listed, and then plotted in one of the four squares. The intention would be to then seek to understand the nature of any risks that fall within the 'unknown unknown' square better, so the risks can then be repositioned into one of the other 'squares'. Although this exercise is fundamentally flawed, as by their very 'nature' the 'unknown unknowns' are the very risk events we are unaware of, and we are unaware that we are unaware of them, thanks to their lack of previous occurrence. Nevertheless, even if this square is left blank, the exercise could still be completed. The intention would be to better understand the nature of all risk events so there is an awareness over which risks are understood in terms of probability but not impact, and which are understood in turns of impact and not probability. Work should then begin to ensure the full nature of those risks are better understood so the risks can ultimately be moved to the 'known known' square. Figure 9.2 explores how this exercise could look in regard to the potential risks associated with the external maintenance of a high-rise building.

**Figure 9.1**   Understanding the nature of risk.

| | | Probability | |
|---|---|---|---|
| | | Known | Unknown |
| Impact | Known | 1. Falls from height<br>2. Damage from passing rioters | 1. Sabotage of security equipment<br>2. Graffiti on facade |
| | Unknown | 1. Lighting strike<br>2. Severe weather event | 1. Crash from vehicle<br>2. Neighbouring building collapse |

**Figure 9.2**   An example of understanding the nature of risk.

The nature of risk is all about understanding and exploring the occurrence of events, against the axis of probability and impact. These categories can help us better understand risk. A better understanding of risk will ultimately lead to an increase in preparations that can be made, and so an increase in the effectiveness of how risk can be dealt with. This practice comes before any risk management or mitigation techniques, as risks cannot be suitably managed or mitigated if they are not fully understood. It could be argued that risks cannot even be suitability retained or transferred to others if they are not fully understood – as doing so could leave one party overexposed. Arguably, and unfortunately, the full understanding of risk, prior to attempts made to 'deal' with it, is not a practice adopted by some organisations and operatives within the construction industry.

## 9.4   Risk in the Built Environment

The built environment can be considered as the design, planning, development, construction, operation, maintenance, refurbishment and ultimate demolition of assets. This definition encompasses the full life cycle of an asset – this asset could be anything from a gas pipe to a 5-star hotel; whilst the risks for each may be different, the approach taken to understand the associated risks should be consistent.

Chaired by Sir John Egan, the UK Government created the Construction Task Force in 1997, with Egan publishing his much-discussed report *Rethinking Construction* in 1998. The report praised the excellence of the construction industry, when it is at its best, but highlighted how generally the industry is underachieving. The report went on to explore this underachievement in terms of low profitability, low investment, and low client satisfaction, and identified five key drivers for the required change. There were committed leadership, a focus on the customer, integrated processes and teams, a quality-driven agenda and commitment to people. These drivers, it was stated, would help achieve a 10% reduction in the time and cost and construction works, and reduce the number of defects by 20%.

In 2009, a further government-backed construction industry report was published by Constructing Excellence, written by Andrew Wolstenholme and titled *Never Waste a Good Crisis – A Challenge to the UK Construction Industry*. The report was written in response to the Egan report to reflect upon the changes the industry had made since the publication of *Rethinking Construction*. Quotes contained within the report from Sir John Egan set the scene for the industry changes that have occurred as they state, 'We could have had a revolution and what we've achieved is a bit of an improvement' and 'I would give the industry 4 out of 10' (Wolstenholme 2009, p. 8). The report also includes a quote from Sir Michael Latham (who wrote the 1994 construction industry report 'Constructing the Team') that states 'What has been achieved? More than I expected but less than I hoped' (Wolstenholme 2009, p. 8). The report goes on to say that whilst the industry has made some progress against some of the goals set in the Egan report, the industry has fallen short more than it has succeeded. In part, any successes the industry had experienced were intrinsically linked to the health of the UK economy, and buoyant times led to financial prosperity and no real motivation to make any changes. During times of economy uncertainty, the problems with the industry are revealed. It is arguably during financially perilous times that industry actors are less likely to be open to change, and instead revert back to previous practices in an attempt to batten down the hatches and weather the passing storms.

With regards to how the construction industry deals with risk, the report states 'scratch beneath the surface and you find many so-called partners still seek to avoid or exploit risk to maximise their own profits, rather than find ways to share risk and collaborate genuinely so that all can profit' (Wolstenholme 2009, p. 8). The conclusions drawn from the extensive industry survey utilised by the report included the actions of the industry in cherry picking desired behaviours rather than wholeheartedly accepting calls for change as organisations continued to focus on serving their own self interests. The survey results also revealed any positive collaborative behaviours did not extend as far as the supply chain, who continued to be the recipients of increased risk despite the opportunities existing for risks to be shared and eliminated together.

The Wolstenholme report also discussed the instinctive risk aversion techniques that are prevalent in every corner of the construction industry and linked this to the often-low margin competitive focus of construction companies. Such behaviours are also not confined to top-tier clients and contractors, with the supply chain operating in the same fashion and passing risks downwards, increasing the chance of disputes and also ultimately putting the risks on the parties who have no other choice but to accept them and not necessarily being those best placed to manage the risks. This is echoed in the report which states, 'Contractors would rather "push" risk down the supply chain than "pull" the opportunities back up' (Wolstenholme 2009, p. 22). The threats of such risk behaviour are that it will prevent and hinder innovative practices from being developed and proposed. If risk consideration and practice does adversely impact organisational behaviour when it comes to embracing innovation, then this can prove problematic in all aspects of the construction life cycle. It can however, be acutely experienced by those with responsibility for maintaining buildings post construction. Firstly, the benefits of any innovative practices and technology adopted in the construction of a building will ultimately be experienced in the operation of that building, even if the innovation resulted in a 'construction phase only' benefit, such as a shorter construction time frame. In this instance, whilst the experience of a building may be the same for the users – such users will ultimately get earlier access to the building than they would have if non-innovative practices and technology had been adopted during the construction phase. Secondly, for the most part, the benefit of innovative technologies is ultimately realised by the owners and end users of a building. From an increased lifespan of mechanical systems to the reduced maintenance required for electrical instruments, innovative technologies and practices are what enable efficiencies to be made and building running costs to be reduced. Generally, a more collaborative and innovation-driven construction phase, adopting a greater awareness of risk and how risks can be shared if correctly managed, could result in greater benefits across the whole life cycle of a building for almost all stakeholders involved.

In 2016, Mark Farmer published *The Farmer Review of the UK Construction Labour Model; Modernise or Die*. In the review, Farmer identified several issues within current construction industry practices that he calls 'symptoms' of the industry's poor performance. These include:

- Low productivity
- Low predictability
- Structural fragmentation
- Leadership fragmentation
- Low margins, adversarial pricing models and financial fragility
- A dysfunctional training and funding model
- Workforce size and demographics
- Lack of collaboration and improvement culture
- Lack of research and development and investment in innovation
- Poor industry image

Whilst some of these issues are more applicable to perhaps the construction phase than the maintenance phase of a building's life cycle, many arguably are applicable to both. These are the present authors' views on how some of Farmer's key symptoms of the construction industry's poor performance can be related to the maintenance of a construction project:

Low predictability

When it comes to the management and maintenance cost of buildings, or any built environment asset, there is a low predictability regarding the time and costs involved. For example, the owner of a residential tower block may be able to calculate a total replacement cost of a certain mechanical element by simply multiplying the total number of that element (say, one per apartment) by the total cost involved in replacement. However, this would only calculate the extreme scenario of having to fully replace all elements. When you consider likelihood of failure, costs of repair, availability of skilled trade professionals to carry out the repairs, and the perhaps intangible cost of disgruntled residents, the predictability of the time, cost and extent of any works reduces dramatically. Unfortunately for the construction industry in general, this low predictability has resulted in an acceptance of failure by both the industry and its clients (Farmer 2016). This is evident in the maintenance aspects of construction assets where preservation against such issues has been dealt with in an aggressive transfer of risk from the client to a facilities management company, which has to then manage the unpredictable nature of the construction industry. Trying to achieve this when tied in to fixed deliverables with a low margin is often very difficult.

Low margins, adversarial pricing models and financial fragility

One of the largest construction industry failures in recent years is arguably that of Carillion. Whilst failures of such magnitude are often a result of a multitude of factors, one of these, in this instance, was the unsustainable pricing of facilities management contracts. Such an approach with constant low margins will undoubtedly win any construction company the contracts they tender for and is potentially a good method of short-term work gain, especially when the contract requirements are fully understood, and any risks have been mitigated and managed effectively. However, when low margins are frequently used as a long-term strategy in procurement, and then the buildings that have to be maintained are done so under contracts where the risks have not been properly understood, and worse still, not been fully understood yet fully adopted as part of the contract requirements, then inevitably problems will arise. Such an approach will only lead to financial fragility and, as in the case with Carillion, when combined with other management and financial issues, it can lead to ultimate organisational failure.

Lack of collaboration and improvement culture

The industry is ultimately competitive and clients and building owners must evidence that they have considered all contractors' tenders and justify any decisions made with regards to awarding contract packages. This, more often that not, comes down to a purely cost-based focus – with the lowest priced tender often the most successful. Whilst there are drives in the industry to change this approach, and it is not adopted by all clients, or those above others in the supply chain, it is a prevalent practice. Such a practice then gets repeated down the supply chain and so the lowest-cost-wins mindset is embedded throughout. This hinders most forms of collaborative practice from occurring. Where any collaborative practices do occur, it is often where each party has an eye on their own commercial interests over the aim of the collaboration and so the ultimate and original intended improvement that the collaboration set out to achieve is never realised. In building maintenance, whilst the immediate opportunity for collaborative behaviour may not be apparent, a facilities management organisation will often outsource and subcontract many of their own contract requirements, and so adopting a lowest price winner procurement model amongst supply chain partners in such circumstances will inevitably lead to a lack of collaboration, or a mistrustful collaboration, where collaboration is essential.

Whilst many of the construction industry reports have discussed risks in some form or another, they all share the opinion that the construction industry, and arguably by extension those involved in the building maintenance sector, all deal with risk in an ineffective manner. Report after report has focused on a different aspect of the industry that is in urgent need of reform and improvement, yet an overlapping and reoccurring theme is how ineffective and adversarial risk practices are within the sector. Risk in the built environment is treated with a reactive approach, a management tool that is used to mask those failures that were preventable, and a way to obfuscate poor planning decisions from external scrutiny. Within the built environment, risk is seen as a negative force, one to be avoided and passed on as quickly and as intact as possible – no one wants to hold risk or retain any element of it, but merely act as a conduit transferring the risks from one party (usually above them in the supply chain) to another party (usually below them in the supply chain).

## 9.5 Risk in Asset Management and Maintenance

The risks the asset management and maintenance sectors are faced with may differ from the risks other sectors face, even sectors within the built environment. Whilst some risks are generic and apply equally across sectors, some risks are more applicable to some sectors than others. Certain risks are also unique to some sectors and may not apply in any form to other sectors or industries. This is why it is argued by some that no two risks are ever the same, even if it is the same risk different parties in different situations face, i.e. the risk of building fire. This risk will have different ramifications depending upon the unique characteristics of each asset and building. Therefore, whilst there can be similar procedures and practices adopted for dealing with risk, the actions required to deal with the risk effectively will be bespoke to the parties involved. Case Study A describes how risk is dealt with in asset management.

Case Study A

The client is procuring a new framework for a series of maintenance and repair projects. During the framework procurement the client shares the proposed framework contract. The contractors who are tendering are given advance notice of the full contract that will be used for all framework projects. As part of the procurement process scoring of the received tenders, the client has told all tenderers that they will be awarded a maximum of 10 points if they accept the contract requirements in full. Any challenges to the contract terms imposed will result in a points deduction; the more contract challenges, the lower the points awarded, and so the greater the risk of the tenderer being unsuccessful. Therefore, if any contactor challenges the contract terms imposed, they will not be on the framework. Whilst some may feel this approach is acceptable and the ethics behind it sound, some may not. Nevertheless, how risk is dealt with in the contract is to push it all to the tenderers. The contract states that all contractors must accept the conditions of each asset, and an asset in a poorer condition than originally expected or advised is not a compensation event. Even though each asset is accompanied by a client report, the client accepts no liability for inaccuracies. Therefore, the contractors each have to accept the fact the only way to guarantee they progress to the next stage of the procurement is by accepting all risk. At the next stage of the procurement exercise, the contractors are then asked to submit their lowest set of costs, with the weighting heavily in favour of the lowest. Essentially, the contractor who is successful in winning the tender for the maintenance of the assets is the one who accepts the maximum risk for the minimum price.

## 9.6 What Is Risk Management?

Once risks have been accepted by an organisation they will then have to be managed. There are many tools and techniques that can be adopted in risk management, but essentially, risk management is the process undertaken to identify all applicable risks, followed by their analysis and evaluation, before suitable action is taken which could include control measures being implemented and alternative solutions being proposed and followed.

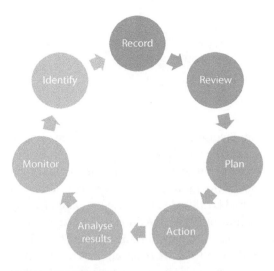

**Figure 9.3** Risk management.

Risk management is an ongoing process whereby the status of each of the identified risks is monitored and updated at regular intervals. In effect, risk management is an ongoing iterative process that is explained in Figure 9.3.

If such an approach is adopted in risk management then, in theory, the majority of risks should be dealt with effectively. The terms 'majority' and 'should' are used here, as even with the best risk management practices in the world, risk is essentially unforeseen occurrences, and people generally have trouble imagining an event that has never occurred before. This is why the global Covid-19 pandemic caught the majority of people off guard, and their first reactions turned out not to be the best to deal with the new risks posed. History and experience should (hopefully) enable future risk management to take account of pandemics with effective plans put in place.

Section 9.7 talks about the two main categories of risk: contract risk and project risk. This section, however, will for now, treat both as largely the same for the sake of risk management procedures and practices, to show how risks of any type can be effectively managed.

What Figure 9.3 essentially allows, is for an understanding to be gained of risk exposure. This is the amount of risk that a current project, building, contract, asset, contractor, maintenance provider, etc. is exposed to at any given moment. Understanding risk exposure is fundamental to industries like insurance, as higher premiums will be charged in line with higher risk exposures. Take, for example, the floods resulting from extreme weather events across the UK; these are often called 'once in a lifetime' but seem to be occurring every few years. For all built environment assets – whether they be residential, commercial, or industrial – if they are situated in low-lying areas and those areas identified as flood plains, then their risk exposure will have risen dramatically in recent times. Fundamentally, this will mean they are at risk of increased flooding and the associated damage this entails, but from an insurance perspective it means such homes and

**Figure 9.4**   Risk exposure.

businesses will have to pay higher and higher insurance premiums. A simple method of understanding and considering risk exposure can be seen in Figure 9.4.

In Figure 9.4, probability is the likelihood an event will occur, with the full ramifications that will be felt if the event does occur. Some ramifications can be considered in financial terms, as in some instances this method is utilised to help calculate and quantify a risk allowance. For example, if you are responsible for the maintenance of an existing school building, and several items of plant are out of warranty but still working fine, there is a risk that some items of plants may fail and so will require repairing/replacing at considerable cost. But if you are asked to compile a repair budget for the forthcoming year, it could be impractical and unfeasible to request a budget to include for the full replacement of all plant that may need replacing, especially when budgets are tight. Table 9.1 lists out the items in this scenario which could be out of warranty and so will require repairing/replacing if they fail.

In this scenario, Table 9.1 is the identification and recording of risks, as per some of the steps in Figure 9.3. This can then be built upon further by following the next steps in Figure 9.3 and adding further analysis and details to the table, as can be seen in Table 9.2.

For this example, crudely, the figure of £20,400.00 has been arrived at for a potential annual maintenance budget. The mathematics behind the calculation can be clearly articulated and the percentages and budgets can be adjusted accordingly to arrive at a new total if required. It is a somewhat easy and straightforward method by which to arrive at a risk exposure total, although on closer inspection faults can be identified in this method. If, for example, the electrical metering systems or HVAC system did fail and a full replacement was required, then the budget would only just cover or not be enough

**Table 9.1**   School Plant requiring repairing/replacing.

| Plant | Full Replacement Cost |
| --- | --- |
| Boiler system | £12,000.00 |
| Design and technology equipment | £3,000.00 |
| Electrical metering systems | £20,000.00 |
| Food technology equipment (student classrooms) | £2,000.00 |
| Food technology equipment (dining hall) | £8,000.00 |
| HVAC system (heating, ventilation, air conditioning) | £42,000.00 |
| IT suite | £3,000.00 |
| Total | £90,000.00 |

**Table 9.2** The risk exposure of school plant requiring repairing/replacing.

| Plant | Full Replacement Cost (Impact) | % Likelihood of Occurrence (Probability) | Risk Exposure (% Likelihood × Impact) |
|---|---|---|---|
| Boiler system | £12,000.00 | 20% | £2,400.00 |
| Design and technology Equipment | £3,000.00 | 40% | £1,200.00 |
| Electrical metering systems | £20,000.00 | 10% | £2,000.00 |
| Food technology equipment (student classrooms) | £2,000.00 | 70% | £1,400.00 |
| Food technology equipment (dining hall) | £8,000.00 | 40% | £3,200.00 |
| HVAC system | £42,000.00 | 20% | £8,400.00 |
| IT suite | £3,000.00 | 60% | £1,800.00 |
| Total | £90,000.00 | | £20,400.00 |

to cover these eventualities. Nevertheless, at its core, the above exercise is a simple one to start to understand the idea of risk management and how risk exposure can be ascertained. Together in a single document, this is the start of a rudimental risk register.

A risk register is simply a document that contains all the risks of a particular project collated in a format that can be monitored and updated. Table 9.2 could easily be classed as a risk register, although depending upon who the key stakeholders are, the financial information may want to be removed. A risk register would also most likely have an owner for each risk, and an action or set of actions that have been or will be taken to mitigate and reduce the risk exposure.

Appreciating risk exposure can also allow the identification of significant risks or map the increasing prevalence and magnitude of risks as they grow and develop overtime. It can also be a worthwhile exercise to allow the categorisation of risks as a base from which to decide whether risks are to be retained or transferred. For example, if the percentage likelihood of the boiler system needing to be replaced is growing and the costs too high to cover in an annual repair budget, increased insurance options could be considered as a way to transfer and mitigate the risk of the boiler costs falling wholly to the building owner. This then falls into the category of risk policy, and it could be that any risk that exceeds a certain financial threshold set is then insured against, and for anything below a certain threshold the risks are retained, as the building owner believes them to be so little that it is not worth paying extra insurance premiums to cover them.

The methods outlined in this chapter all revolve around the identification of risks and risk exposure from a cost perspective. This in, in fact, one of the prevalent themes in countless construction reports that have been published. Risk is often identified in terms of cost, with the financial implications of risk ascertained. Even when risk is discussed in terms of time or quality, this is then often translated to financial metrics, i.e. a delay of one week would cost £10,000.00, or the cost of replacing poor quality materials is £5,000.00. Cost, it appears, is the nature of risk management in the construction and maintenance industry.

## 9.7 The Nature of Risk Management

Whilst in this chapter risk has been broadly discussed as a single entity, it can actually be considered from two broad perspectives. The first is project risk. These are the risks most would associate with maintenance and repair projects, for example the risk that an operative would be injured during the carrying out of any works required. The second category of risk that could be used is contract risk. This refers to when one party who is in a contractual agreement is left exposed to any risk events that occur. This exposure could be planned as part of the contractual allocation process, whereby responsibilities are allocated amongst the parties involved in the contract, with one party aware of their full responsibilities and requirements. Alternatively, it could be a situation whereby the full contractual responsibilities of one party are not fully known at the outset, either from the party not understanding their responsibilities, or them being 'buried' in the contract clauses purposefully by the other party.

If we go back to the allocation of responsibilities under the contract, this is often a misleading concept. In most cases, contract responsibilities are not negotiated and allocated between parties, but simply imposed from the higher supply chain partner upon those lower in the supply chain. In this sense, risk events are not simply shared, but forced. One party is therefore compelled to accept the events the other party deems they should be responsible for, and if such events then occur, then the party with the responsibility (those lower in the supply chain) will have to bear the risks realised.

Risk management is therefore the process of dealing with such risks, both those that may occur during the project and those contractual risks dictated and mandated by those higher up in the supply chain. These two types of risk may not, however, appear on the same risk register. Once again, it comes down to those stakeholders that will be viewing the risk register on a regular basis. If it is a shared project risk register of all the health and safety hazards encountered on-site, it would not be in the best commercial interests of any party to have their contractual risk exposure a well-known topic of conversation, as in some instances one party may be exposed to a certain risk in a contract, but the other party to the contract may actually be unaware of the details of such risks.

## 9.8 Risk Management in Asset Management and Maintenance

In asset management and building maintenance, risk management pertains to both project risks and contract risks. It is important this is understood, and any risk registers developed with the intended stakeholders in mind so they can form a useful document that is regularly reviewed and updated, and not something created at the start of a project and then placed in a dusty drawer, only retrieved when a risk is realised.

Often, risk registers are the tool adopted to evidence risk management, but they are utilised to review what has actually occurred on a project, or what the next course of action is. A risk register used correctly is a proactive tool regularly reviewed to identify any new risks and update the likelihood of existing risks. Whether these risks are increasing or reducing in likelihood will help recalculate the current levels of risk exposure. In the maintenance of buildings some of the most commonly occurring risk events include:

- Falls from height – Whether involving in-house employees or subcontracted staff, people or objects falling from height are the most common occurrences of a risk event being realised. Simple techniques and methods have been employed over recent years to reduce both the likelihood of an event occurring and the impact if it does occur. For example:
  ○ Reduce the likelihood – Have all operatives use stable raised platforms to gain access when working at height instead of using ladders (or worse). Ensure all tools used at height are attached to the operative using lanyards. Safeguard those working and passing the area where works are being undertaken, by having the area cordoned off for the duration of the works and all access restricted.
  ○ Reduce the impact – Ground mats could be used to surround the access platform so that if an operative does fall, the likelihood of serious injury will be reduced. Ensure only the minimum number of tools are taken up on the platform at any one time. Safeguard those who have to work underneath or nearby by having strict rules in place with regards to personal protective equipment (PPE), such as hard hats.
- Disturbing asbestos – As a very common building component for many years, asbestos can be found in all buildings that were built prior to the year 2000 in the UK. And so, with a considerable amount of residential and commercial buildings currently in the UK stock that were built before this period, asbestos is an issue of considerable consequence to all those working in building maintenance. However, once again, techniques can be employed to both reduce the likelihood and impact of serious harm. For example:
  ○ Reduce the likelihood – The law does not require asbestos be removed, but individuals are required to manage it. Having up-to-date and accurate building condition surveys will serve to inform those responsible for planning work as to whether asbestos is present in an area. If so, the decision may be made to encase as opposed to disturb when attempting to remove.
  ○ Reduce the impact – If asbestos is to be removed, then having specially trained operatives with the correct PPE and means to remove and dispose of it safely, ideally working in a well-ventilated area, will be help reduce the impact of immediate and future harm being caused.

In building and asset management and maintenance all risks are usually dealt with by a hierarchy of control measures. These can be seen in Figure 9.5.

This fundamentally means once a risk register is created, and full extent of risk exposure has been identified, the hierarchy of control measures can be introduced in an attempt to mitigate and minimise the risk exposure of each risk activity.

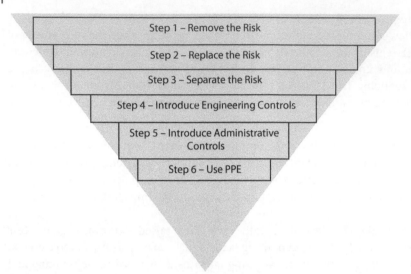

**Figure 9.5** Hierarchy of control measures for risk.

Step 1 – Remove the Risk

In the first instance, if the risk activity can be removed, then it should be. This step is often confusingly interpreted by some organisations and individuals to fuel their belief that risks can be transferred to other parties to deal with, i.e. a client believes the risk of finding asbestos can be removed by simply stating in a contract that all asbestos and building condition risk is to be the responsibility of the organisation contracted to maintain the building. Now, the 'confusion around interpretation' of this step could be greeted by sceptics as somewhat purposeful, so clients and those above others in the supply chain have legitimacy to 'remove' risks from themselves by transferring them to others. This, in the opinion of the authors, is incorrect. Whether actioned by an accidental incorrect interpretation of step 1, or if the first step is purposefully repositioned to suit a party's actions, is somewhat irrespective. The 'removal' of risk in this step simply refers to the risk being removed in its entirety. Therefore if, for example, a risk exists in upgrading a building's insulation, as asbestos was used in the construction of the building for loose fill insulation, the removal of this risk would be to leave the asbestos insulation in situ, and simply add a form of internal wall or external wall insulation. Whilst the risk would ultimately still exist for those who will one day demolish the building, for the purposes of maintenance, the risk of disturbing loose fill asbestos will be massively removed. If removal is not an option, the next stage is to replace the risk with an activity or method that is ultimately less risky (has a lower risk exposure).

Step 2 – Replace the Risk

Replacing a risk is effectively substituting one action or activity that achieves an outcome, for a different action or activity that achieves the same, or a very similar outcome. For example, in asset maintenance, if there is a gas pipe that requires repairing through grit blasting and painting activities, but the pipe is buried underground, excavation with plant and machinery may run the risk of damaging the pipe. This activity could therefore be replaced with hand digging, that although will take longer, will reduce the risk of accidental pipe damage. If the risk cannot be reduced through replacement, then separating the risk could be considered.

Step 3 – Separate the Risk

This is where the risk or risks are separated from activity as best they can be. For example, if we have a gas pipe over a live railway line that needs repairing, then there are a multitude of risks involved. The next step would be to separate out any risks from the activity that can be separated. For example, the activities involving carrying out the full repair (spending a week working from an access scaffold above the train line) could be substituted with activities that still achieve the ultimate aim of repairing the pipe. A real risk here could be collision with passing trains, so therefore the obvious separation would be to only conduct any repair works when trains are not passing nearby, by working at night, or reaching an agreement with the rail operator to have trains suspended for a time to allow nearby repair work to be conducted. Other examples of risk separation could be the separation of groups who could be exposed to the activities being carried out. For example, if works are being completed in a public setting, then having physical barriers to prevent members of the public from coming within close proximity of the works would separate them from the risk of accidents occurring. Where risks cannot be separated, then engineering controls should always be introduced.

Step 4 – Introduce Engineering Controls

Similar to step 3, the introduction of engineering controls effectively separates some of the risks involved in activities. These can be at a more practical and site-based operative level and are concerned with controlling and safeguarding the exposure to sources of risks and hazards. For example, mechanical guards and vacuum equipment attached to plant as well as localised barriers and safety screens. For example, if repairs to a hospital electrical system are required, and it is impractical to separate the risk by closing a corridor of the hospital off whilst the works are being undertaken, then utilising a small moveable screen to cover the works undertaken and any exposed wiring from easy public view may be an appropriate engineering control. Examples could also include vacuum devices to prevent the spread of dust when cutting, and acoustic blankets to minimise the spread of noise.

Step 5 – Introduce Administrative Controls

Where activities have to be undertaken, and all other steps have been acknowledged and followed, administrative controls should also be introduced. These include ensuring all staff are suitably trained with all skills, qualifications, and competencies up to date. Organisational safety processes also need to be in place, such as quality management systems and safety management systems to ensure good practices and procedures exist and everyone is aware of the actions that are required at every step of planning and conducting of works. Once all aspects of the hierarchy of controls ha been followed, and all operatives are trained and carrying out activities that must happen in a manner that poses the least amount of risk to all stakeholders, then and only then should PPE be considered.

<u>Step 6 – Use PPE</u>
PPE is of the upmost importance. Considering it last should not relegate its importance as a control method to reduce risk, but merely illustrate how a multitude of other factors should be considered prior to carrying out any repair and maintenance works on buildings and assets. Only when all other control measures have been considered and it is decided that a certain activity conducted in a certain manner must be completed should PPE then be discussed. PPE will help minimise the impacts of any risks encountered; for example, for an operative working at height, having ground protection mats will cushion the impact of any fall. Hard hats will reduce the physical impact of a falling object and high visibility clothing will ensure the operative is seen more easily which will help when they are operating in environments with heavy traffic.

## 9.9 How Is Risk Classified?

One frequent occurrence when it comes to risk that may have started to become prevalent during this chapter, is that risk is ultimately considered in financial terms. Monetary metrics are the ultimate currency of risk. Consider the risk that inclement weather may prevent external building maintenance from proceeding as planned. The risk is that it will either cost more to go ahead with the works as additional façade and operative protection measures will need to be utilised, or the works will be delayed, which will mean the condition of the building will be worse when the repairs and maintenance is eventually undertaken, which will result in higher costs being incurred. Even seemingly small risk control measures can be considered from a long-term financial position. If we take step 6 (the use of PPE) in the hierarchy of control measures, then any tender that is ever returned to complete works will have taken into consideration the cost of supplying each operative with the required PPE, and this will have been built into the tender price. That is not to say cutting health and safety corners should be carried out to save money in the short term, as operative protection and welfare is of the upmost importance, but just that all risk control measures are more often than not classified in terms of monetary metrics. This approach is adopted throughout the supply chain with each organisation having a risk budget and allowance and building into their returned tenders (wherever possible) a financial amount to be used in the eventuality that any risk is realised.

Even when seemingly non-financial metrics are used, for example measuring the amount of $CO_2$ produced, whilst the metric might be categorised in tonnes, a financial penalty will be associated with exceeding any limit. The link between non-financial and financial metrics can also be witnessed in the number of serious or fatal injuries occurred that are reported under RIDDOR (The Reporting of Injuries, Diseases and Dangerous Occurrences Regulations 2013), as such events will inevitably be considered in financial terms. For example, if a death was to occur, this would result in a fine against the company, and so the company has a tangible financial figure against which the risk of not caring for their workforce can be compared (as well as the potential loss of current and future contracts and operatives). This is, of course, taking a less compassionate view, as obviously the vast majority of organisations view the safety of their workforce and any subcontractor workforce as above any cost, and of paramount importance. The argument is merely that in all cases, in addition to the social and environmental focus of risk, monetary metrics will always be in consideration. Somewhat regardless of the classification of risk, monetary metrics will also be strongly associated with how risk is discussed.

This, however, is not a bad thing, even if some feel it may detract away from the purpose of risk management. By assigning a monetary metric, someone somewhere will sit up and take notice. They will not want to spend figures associated with risks if they are realised, and so in comparison mitigation costs are often much lower. Therefore, if risk management comes down to financial control, mitigation measures will often win on a lowest cost basis, as if the risk is realised, it could leave all parties exposed to bear the brunt of much higher costs. Although, conversely, this does often mean that some risks are deemed lower and so less important if they are realised. Often, where budgets are stretched this may result in the allocation of funds to some risk prevention methods over others. In such circumstances, social risks are often overlooked, especially around the growing concept of social value. If the types and amounts of social value are not achieved, often this carries no financial burden for those organisations tasked with the social value delivery. This situation can and will change once social value gains more traction as a key procurement criterion with greater weighting in comparison to the triumvirate of time, cost, and quality. At present, however, not achieving social value is illustrative of a risk that will be borne if it came down to not achieving that over another risk event, such as ensuring works are repaired in the contractually agreed time frame.

## 9.10   Risk Events in Building Maintenance and Asset Management

If an event is deemed a risk event (in that not achieving the event will result in a risk being realised) then it needs to be identified and planned for as the above sections have shown. However, there are some risk events that are well known in building maintenance and asset management. Having an awareness of these common events can help proactive planning and ensure the hierarchy of controls is visited well before any works are close to starting, or even being procured. A proactive and effective building manager or facilities management company will be aware of such events generally, and then simply tailor their knowledge and processes to any bespoke factors applicable to each building or asset. Some of these common events are shown in Table 9.3.

**Table 9.3** Common risk events.

| Risk Event | Occurrence in Building Maintenance/Asset Management |
|---|---|
| Falls from height | When working on assets that are above the ground, or maintaining building elements that are at height, there is risk an operative may fall, sustaining injury. This could be cleaning first floor windows on commercial property or falling whilst installing or repairing cladding on a high-rise building. Ensuring the work is necessary to be carried out at that height in the first place is the best place to start and if it is, ensuring access platforms and not ladders are used when working at height is required could be an idea, and have all operatives attached to the platform to help prevent accidental falls. |
| Slips, trips, and falls | Whilst repair works are carried out in busy premises, trailing cables could lead to trips and falls of both operatives and those stakeholders present (i.e. customers/residents, etc.). Conducting any works when others are not present is one method that could be employed, with clear segregation of areas if not, and the use of signs and battery powered devices wherever possible. Operatives should also seldom work alone, and so if any accidental spillages should happen, one operative can engage in cleaning the area whilst the other continues the work. |
| Inadequate equipment | Faulty equipment can be a danger to users and those around at the time any malfunction may occur. Ensuring all equipment is regularly maintained and PATs (portable appliance tests) are conducted will go some way to minimising faulty equipment. But, equally, ensuring training is up to date and equipment is used correctly and for the correct purposes will also help minimise any unnecessary accidents. |
| Child labour | If supply chains are long, it runs the risk of oversight being lost, especially if they cross international lines and multiple components are involved. Nevertheless, child labour is a very real issue that should not be ignored through ignorance or oversight. Supply chains should be reviewed regularly, and standards upheld if relationships are to continue. The standards to enter an approved supplier list should extend more than merely lowest cost and comprehensively consider a range of factors. |
| Forced labour | Subcontractor labour, if elements of any required materials are completed off-site, could be linked to forced labour practices either directly or indirectly. Great care should be taken to analyse full supply chains to ensure quality and standards (and legal requirements) are upheld at every stage of a supply chain. Whilst it is easier for those higher up in the supply chain to govern the behaviours of those below, any supplier and contractor does have a moral obligation to ensure any party they do business with (below or above them in the supply chain hierarchy) behave ethically. |
| Poor finished quality | This risk is that any work that is undertaken does need meet the required quality standards once complete. This can lead to an increase in prevalence and severity of other risks – such as a lower valuation of a building or asset, or lower rent yields as the quality of the whole building is not high, or increased risk of injury to users. For example, if a door is repaired to a poor quality in a shared corridor of a residential tower block, then this door could lead to an injury should it fail to open or close as expected by a resident. |

*(Continued)*

**Table 9.3** (*Continued*)

| Risk Event | Occurrence in Building Maintenance/Asset Management |
| --- | --- |
| Works overrunning | If repair works have been procured, then they will almost always come with a time frame in which they need to be completed. Often this time frame will then be used for additional planning. For example, if a school needs to have its roof replaced, then non-urgent work will be scheduled for the summer holidays. This will have both the ideal (in theory) weather conditions and a long period of time in which the school is not occupied by students for the works to be completed. However, a risk would be that roof repair works would then overrun and not be complete in time for the start of the next term. This would be a risk to students and staff, as operatives will be working on a roof potentially above a classroom or overlooking a playground and any associated impacts of any risk events will be increased. |
| Material shortages | A shortage of material availability can prove problematic when urgent repairs are required. Space and budgetary concerns may prevent the large scale stock piling of surplus materials when they are not needed and so this is a real risk event that, depending upon the materials required, could be hugely problematic in building maintenance. It is best practice to be aware of all potential materials that may be required for future repairs and ensure either a surplus is retained where possible, or sourced with as much notice as possible by ensuring a high awareness of the condition of all plant and equipment. |
| Staff turnover | Having a stable workforce may not be a requirement of organisational success, but it does help. The consistency of delivery, as well the time and cost savings of having to train operatives up from scratch to understand methods and techniques of company working practices, are some major impacts felt by organisations who have a high staff turnover. It also potentially reveals bad working practices, and organisations that have low staff turnover will undoubtedly advertise such facts and so it could prove to be a disadvantage to those who have high staff turnover when they are being compared by either potential clients or potentially employees. This is a risk if a professional knows the maintenance plans of a building or asset and is familiar with the client/building/asset/systems in place. If they leave and a new professional takes over, the initial period where steep learning is involved can prove to have an increased risk of events occurring not as planned and so the risks being realised. |
| Changing legal obligations | It is of the upmost importance that those involved in the building maintenance sector are aware of all current and upcoming legal changes. To offer clients advice that is out of date, or to instruct works to be completed that do not comply with the latest guidelines and requirements can be potentially dangerous and is easily avoided with both proactive and reactive ongoing training and continuing professional development. |

## 9.11 The Consequences of Risk Events

If a risk event was to occur and the risk realised, they can be broadly classified into one of the following categories: reputational, financial, operational, relationship, operational or legal.

Reputational
If a risk event is realised then depending upon the severity and those stakeholder groups who are impacted, the realisation of the risk event could lead to reputational damage for

the organisation. For example, a building maintenance company that has a history of poor health and safety management of its operatives may suffer from a bad reputation when it comes to its workforce well-being (and rightly so). Also, if a building mainte- nance company focuses purely on the lowest cost and does not consider finished quality of their works as highly as making the maximum profit, this could also negatively impact their reputation with current and future clients and building owners who could poten- tially view the building maintenance company as one who would not care for their building over the long term, even if they do return the lowest tenders during procure- ment exercises.

### Financial
Such reputational damage could then manifest itself in poor financial returns. The build- ing maintenance company that suffers from a poor reputation for quality may find it increasingly difficult to secure future contracts. We have covered how financial metrics are the underlying currency of risk, and so the realisation of any risk event will ultimately take a financial toll on a company involved. In addition to loss of work, the financial implications could also be felt through fines and the need to pay staff higher salaries to work for that particular organisation over competitors.

### Operational
If a building maintenance company does not treat its staff well, then good staff will ulti- mately leave to secure positions elsewhere. The staff that remain could become disillu- sioned with the company's objectives and so not believe in the work they are doing or have trust in the company to look after their best interests. High staff turnover will also result from such a culture and so a company could have a predominately 'new' work- force for each contract they undertake. This would increase the likelihood of risk events occurring, as familiarity with company practices and contract requirements are still being embedded.

### Relationship
Trust in relationships with all stakeholders is often key to the success of a building mainte- nance company. Clients need trust that the company can deliver what is agreed to the qual- ity and standards expected. The public need trust that works maintained by the company are carried out to a high standard with due regards to the welfare of those in proximity during the undertaking of the works and those who will be using/passing the area(s) after- wards. The company's own workforce also needs to have trust that their well-being is con- sidered, and they are working for a company that is supportive of their immediate and future development needs. Without trust in place all these relationships can be broken. If a risk event is realised, then trust can suffer if it is believed the company could have pre- vented such an event from occurring or did not properly prepare for its likelihood or impact.

### Legal
Whilst not all risk events will have legal consequences, there are many applicable to building maintenance companies that do. Ensuring correct quality standards are achieved,

all required health and safety provisions and regulations are adhered to, and relevant legislative requirements are met is of paramount importance. A failure to do so, whilst having wider ranging ramifications that can fall into any or all of the earlier categories discussed, can also fall within a legal category that can result in company fines, as well as sanctions for those involved and those in charge of the company itself. Legal requirements are ultimately there for the protection of all stakeholders, and so an up-to-date awareness of all legal obligations is a prerequisite for all those who work in building maintenance and asset management.

## 9.12 Proactive and Reactive Risk Management

The tools and techniques discussed in this chapter are effective risk management methods and can help promote discussions around risk which will ultimately lead to a greater appreciation of risk events. However, one issue that needs to be discussed is the two different times at which risk management, and the associated discussions, can occur.

Proactive risk management
Risk management can be a proactive tool in the arsenal of a building maintenance professional. Discussions pertaining to risk events can be held frequently, with risks assessed and risk registers updated accordingly. Whilst this does occur in practice, it is all too often not completed adequately and by enough professionals.

Reactive risk management
Reactive risk management is when a risk event has occurred and then the process of risk management truly begins. Whilst in some cases some risk management practices will have been held, they are revealed, once the risk event has occurred, to be unsatisfactory and so the inadequacies are addressed at the same time the risk ramifications are being felt.

Whilst from the above, and the majority of this chapter, it may appear that proactive risk management is the best approach to adopt (and you would be correct in thinking this), reactive risk management practices also have their benefits. Both proactive and reactive risk management approaches should be adopted for their respective strengths to help all professionals form a fuller understanding of risks. Proactive approaches allow risk events to be understood before they occur, and so their likelihood and impact both reduced, and contingency plans to be put in place where suitable. Reactive approaches allow risk events to be analysed, and as long as such approaches are not used to intentionally obfuscate any information pertaining to the risk event for the protection of certain parties' involvement (or lack of), such practices can serve to help better understand why a risk event occurs, and how good the initial proactive risk planning strategies were. Such lessons can then be learnt to better inform future proactive risk practices. This relationship is illustrated in Figure 9.6.

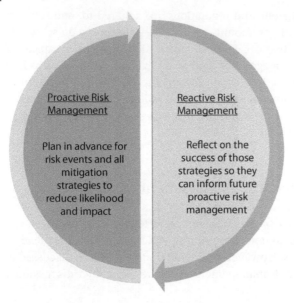

**Figure 9.6** Proactive and reactive risk management.

## 9.13 Procurement Risk

We have discussed many risk management and mitigation techniques, as well as wider industry practices. However, one of the main risk management practices adopted by many in the building maintenance and asset management industry is to simply transfer the risk during procurement. Whilst someone somewhere (lower in the supply chain) will have to manage the risk and will bear the exposure of the risk event being realised, the parties higher up in the supply chain will feel they are protected against the risk event being realised.

If we take an example of a facilities management contract for a multistorey office build-ing, the owner may not want to retain any risk for maintenance of the building and so will procure a building maintenance company. During this procurement process all the risks the client does not want to retain will be passed to the successful building maintenance company as part of the agreed contractual requirements. The building maintenance com-pany will then do the same with any subcontractor they in turn procure to complete aspects and elements of the works. If the company directly employs its staff, they will retain many of the risks themselves. Procurement, however, remains the single largest exercise by which risks are contractually transferred to another party.

Therefore, any professional involved in any procurement and tender exercise must be aware that what is happening at the heart of the exercise is the transferring of risk for a set amount of money. The understanding of this risk is therefore of paramount importance to ensuring all returned tenders suitably protect the party who will be adopting the risks going forward. Whilst this is a tightrope exercise when compiling and submitting prices, a focus on lowest cost at the expense of fully understanding the risks involved can leave a party exposed should those risk events occur, and the risks be realised.

## 9.14   Why Risk Events Still Happen

As we have covered throughout this chapter, with the importance of understanding risk and having suitable risk management strategies in place, some may be thinking, why would any party even act differently? Why would risk events fail to be understood, why would proactive risk management not occur and why would parties be hit by the serious ramifications of risk events they have hardly considered? Table 9.4 highlights a few of the reasons that go some way towards answering some of these questions:

**Table 9.4**   The reason risk events still occur.

| Reason risk events occur | Description |
| --- | --- |
| Lack of knowledge, education, and experience | Those transferring the risks often will do so due to lack of knowledge of the full ramifications of the risks. For example, they may think that by transferring all the risks they are protected, without realising that the risk event may now be more likely to occur and they will still face some of the ramifications. Consider a school that transfers all risks for the maintenance of the external recreation areas to the lowest priced contractor, but the contractor then fails to complete the requirements to a good enough quality. Yes, the school can claim against the contractor for breach of contract, but in the meantime, they have no useable external areas and so will suffer in the short term due to their lack of knowledge regarding transferring risk. This can equally occur due to a failure in advice from any consultants they have engaged. |
| The client already thinks they do this | In many situations, the client (or those with responsibility for the risk) believe they are dealing with it at present, but actually due to a communication failure within their organisation the risk (although acknowledged) has no current plans in place to manage it. |
| It is happening, but not correctly | Similarly, those with responsibility for the risk may have practices in place to manage the risk, but these practices may not go far enough to adequately protect against the likelihood or severity if the risk is realised. Again, this can stem from a lack of understanding or poor advice received. An example could be a building maintenance company that accepts a facilities management contract and underestimate the amount of call-outs they will receive, and so will then have to recruit additional staff to fulfil the contract requirements. |
| Overly optimistic / underestimating the risk | Those with responsibility for managing risks can occasionally be overly optimistic of the likelihood the risk will occur, or the impact of it should it occur. This can lead to a failure to adequately prepare for the most likely outcome, or the worse case outcome, and so if these outcomes were to be realised, an overly optimistic company could be in danger of being underprepared for the ramifications of the risk event. |
| Believe it's someone else's responsibility | One common factor with agreeing contractual requirements is fully understanding the obligations of each party. One party may sign up to the contract not fully understanding what is required of them, and so be immediately unprepared for any risks they may have responsibility for. This is quite common, as repeatedly industry surveys reveal contracts are signed without one party having a full understanding of what their expected deliverables are. |

*(Continued)*

**Table 9.4** *(Continued)*

| Reason risk events occur | Description |
| --- | --- |
| Fear of collaboration | One barrier that has often held back some risk management practices is one party's fear of collaboration with another. As risks can be monetised, a monetary value can ultimately be attributed to the practices that mitigate any risks. Such practices then afford the party who developed them to be in a position to lower their price on returned tenders, and so by sharing these practices they are effectively allowing competitors to have access to their own best practices. Whilst this is true, it is only by adopting a collaborative approach with all parties that risks can be full understood and managed. The client, or those with responsibility for risk allocation need to lead by best practice and encourage the sharing equal practices from all companies involved and help build long-term collaborative and framework relationships. |

## 9.15 Conclusion

This chapter introduces the concept of risk. The nature of risk and how it impacts the practices of built environment professionals is discussed and the risk management concepts and strategies used to better understand risk are highlighted. Risk management is an ongoing and evolving process that needs to be discussed at regular intervals amongst the entire project team, and can have serious implications for the repair and maintenance of buildings. Historically, risk has been a fearful concept in the industry. However, despite all the negative connotations of risk within building and asset maintenance and the wider construction and built environment industry, risk can be an opportunity. How risk is dealt with has traditionally been through aggressive and downwards transfers with often too little thought gone into who is best placed to manage the risk and how the risk(s) can be dealt with effectively or even managed out at earlier stages of the maintenance process. However, this is not always the case, and more and more examples of best practice are being developed and change is becoming increasingly evident. The first step is to truly understand risk, and then become familiar with risk management. How risks can be effectively and efficiently managed is ultimately at the heart of how efficient and effective the maintenance and repair of buildings and assets are. A comprehensive consideration, understanding, and actioning of risk will lead to an improvement across how works are delivered and is often key to the success and failure of many repair and maintenance projects. Put simply, by having a greater understanding and appreciation of risk, the behaviour and practices of built environment professionals should improve, and the industry will be a fairer, more effective and more productive place to operate.

## References

Egan, J. (1998). *Rethinking construction: the report of the Construction Task Force to the Deputy Prime Minister, John Prescott, on the scope for improving the quality and efficiency of UK construction.*

Farmer, M. (2016). *The Farmer Review of the UK Construction Labour Model: Modernise or Die, Time to Decide the Industry's Future.* Construction Leadership Council.

Latham, M. (1994). *Constructing the Team. Joint review of procurement and contractual arrangements in the United Kingdom construction industry.* Crown Copyright.

RICS (2012). *RICS New Rules of Measurement; Order of Cost Estimating and Cost Planning for Capital Building Works*, 2nd ed. RICS. Coventry.

Wolstenholme, A. (2009). *Never waste a good crisis. A review of progress since rethinking construction and thoughts for our future.* Constructing Excellence.

# 10

# Managing the Maintenance Process

## 10.1 Introduction

It is important to have an appreciation of the maintenance process, and for the key management requirements to be understood. It is only through such an appreciation can any maintenance project be managed effectively and the intended targets achieved. The effective management of maintenance works can also lead to reduced health and safety accidents and increased building performance. Therefore, the following chapter gives an overview of how to manage the maintenance process. Key areas of importance are covered, including how to plan for maintenance works and the distinctions of proactive (scheduled) and reactive (corrective) maintenance. The importance of maintenance schedules and how they can form the basis of wider programmes of work is also discussed, as are the key benefits of programming and planning and the key attributes required from those collaborating in the creation and development of a building maintenance programme. The importance of inspecting works at the earliest opportunity is also included, as are the key considerations to be made regarding site and task constraints. An overview of prevalent and successful health and safety initiatives is provided and an introduction to and discussion of the benefits of adopting a soft-landing approach discussed. Finally, the current and potentially future practices of the building maintenance professional's role is discussed, and how by engaging them at an earlier stage in the construction process can lead to a more successful operating of a building.

## 10.2 How to Manage Building Maintenance

Having an awareness of all the principles and practices of building maintenance is a good start, but actually undertaking any maintenance works can prove more difficult than initially envisaged. Most construction projects and building maintenance contracts reveal many more issues for the professionals involved that then need to be resolved in addition to the 'planned' challenges stakeholders had awareness of prior to starting the works. Whilst to some degree this will always happen, as any works undertaken in the built environment are subject to unknown elements and outside influence of wider geopolitical factors (i.e. severe weather events causing unexpected damage, Brexit resulting in labour shortages, and COVID-19 resulting in materials shortages). Nevertheless, successful

*Introduction to Built Asset Management*, First Edition. Dr Anthony Higham, Dr Jason Challender, and Dr Greg Watts.
© 2022 John Wiley & Sons Ltd. Published 2022 by John Wiley & Sons Ltd.

maintenance management of buildings and assets needs to be understood from both a proactive and reactive viewpoint.

Reactively, building maintenance professionals need to be adaptive and flexible. They need to be aware that unexpected challenges will occur and so whilst the exact nature of such challenges will be unknown at the outset of any works, such professionals will need the skill sets to suitably deal with such challenges. There does need to be a clear distinction between what are and are not reactive problems. Too often problems are dealt with reactively when in fact with consideration and accurate planning, such problems would have been revealed to be those that could have been identified before they occurred and then their impacts minimised through proactive management of building maintenance. This a key distinction that building maintenance professionals need to be aware of. Whilst both proactive and reactive problems are ultimately dealt with in the same manner and lessons need to be learnt from both type of problems experienced, being aware of how problems can be proactively identified and dealt with is key to building maintenance success.

Proactive building maintenance is where potential issues are identified prior to their possible occurrence. This involves a collaborative approach between all stakeholders with open and honest discussions of the project details and the sharing of knowledge and experience from past lessons learnt. Ultimately, successful proactive practices involve effective planning for building maintenance.

## 10.3 Planning for Building Maintenance

Effective planning practices are essential for ensuring buildings are maintained effectively and efficiently. Some maintenance can be planned in advance, for example servicing of plant, equipment, and machinery. Whilst some plant, equipment, and machinery may not have an annual service requirement, there will need to be ongoing and periodic inspections to ensure they are operating as intended and so these should also be planned for to ensure they occur at frequent intervals. Embracing the principles of planning is as much about record keeping as it is about forecasting. Accurate and well-executed plans can serve as a record of what happened and when, and so if issues do arise, it can be quickly understood when the last inspection or service took place. When forecasting maintenance work, contingency planning can also allow for periods when key equipment will be down to be identified. This will allow such equipment to be replaced on a temporary basis, allowing production and efficiencies to be maintained. Ultimately, building maintenance should be planned for on a proactive basis. This will prevent the number of reactive maintenance events that are required to be undertaken and ensure any plant, equipment, and machinery operates at optimum capacity.

## 10.4 Proactive Maintenance

Proactive, planned, or even preventative maintenance should be a priority of all building maintenance professionals. It is essential to ensuring the smooth-running operations of buildings, facilities and assets. Its key advantage is that it allows the early identification of issues that could then prove to be serious problems later down the line if left undiagnosed. Often, such issues will never be diagnosed until the full force of the problem is realised and at which point the full ramifications will be felt. At this stage a problem may be costly to fix and take a substantial amount of time to rectify, which could mean further problems are then encountered, resulting in additional costs and further time delays. For example, if pipework is not inspected on a regular basis a small water leak may occur undetected. This could then develop further, and water damage could increase and spread beyond the initially impacted areas. Such damage may then spread internally and lead to further damage to other areas such as paintwork and carpets which ultimately, if unidentified and therefore unrectified, could lead to substantial damage if the water leak encounters an electrical supply. Regular, planned and 'proactive' maintenance would help identify such issues promptly, preventing further problems from developing. Detailed proactive maintenance could prevent the issue from occurring in the first place, if as part of a regular proactive maintenance schedule of areas and fixings that may be subject to leaks are checked, tightened, and replaced as necessary. Whilst proactive maintenance is an effective key tool in the arsenal of a building maintenance professional, and if completed correctly will definitely reduce the likelihood and severity of many issues arising, an element of reactive maintenance will always be required.

## 10.5 Reactive Maintenance

Reactive maintenance can also be described as unplanned or unscheduled maintenance. A certain amount of reactive maintenance will always be necessary, and so it is not the failure of any building maintenance professional or proactive maintenance plan if reactive maintenance occurs. Indeed, reactive maintenance is inevitable and so must always be 'planned for'. This is not a contradiction in terms, but in fact a sensible approach to risk management (as covered in more detail in Chapter 9).

Firstly, reactive maintenance will occur if an item becomes damaged or is not functioning at its full or required capacity and so needs to be repaired or replaced. The maintenance of this item will then be classed as reactive. If a proactive maintenance schedule is in operation, reactive maintenance will still occur for a variety of reasons. For example, severe inclement weather could lead to a nearby tree damaging the drain water system on a building, which could then lead to the rainwater collected on the roof entering the building and causing damage to electrical systems and partition walls and decoration. A vehicle from a nearby road or the building's car park may reverse into a bollard or leave the road and collide with a wall that encases soft landscaping, or vandalism could occur in the form of spray-painted graffiti on an external wall. This would need removing, and potentially enhanced security measures installed, such as perimeter fencing and/or a CCTV system in that area.

One of the benefits of having a proactive maintenance schedule in place is that any reactive maintenance requirements will be minimised. The three brief examples described above can never really be fully prepared for on a proactive maintenance schedule, but plenty of other maintenance examples could be. For example, plant will need to be serviced, and equipment will need to be tested and checked on a regular basis to ensure it still works effectively. A failure to do so will inevitably result in an increased prevalence of reactive maintenance events occurring. In any comprehensive proactive maintenance schedule, however, a degree of consideration should always be given to any reactive maintenance issues, even if they cannot be fully envisaged at the time a project or contract commences. Time may not be allocated, as it would be impossible by their very nature to predict when unplanned maintenance may occur, but costs can be allocated to cover any reactive maintenance requirements so that if and when such events do occur, the financial resources are available to quickly action any maintenance requirements.

## 10.6 Maintenance Schedules and Budgets

Similar to the manner in which risk registers should be used to arrive at a risk allowance to cover the realisation of the identified risks, proactive and planned maintenance schedules should be used, amongst all their many positive purposes, to arrive at a maintenance budget. Some items will be easier to attribute costs to; for example, long-standing, pre-planned annual equipment maintenance will have costs agreed well in advance, and framework agreements may even be set up covering a period of five years or longer with dates and costs for maintenance pre-agreed. Therefore, the allocation of budgetary funds can be easy to justify. Unplanned or reactive maintenance should always be considered, however, when allocating budgets for maintenance. Whilst, as discussed earlier, a thorough and robust proactive maintenance schedule will reduce the likelihood and impact of reactive maintenance events, they may still occur (who can plan for a stakeholder's car hitting a wall?), and so budgets will need to be available in order to conduct such maintenance work as and when required. If such events, or at least the occurrence of wider external events, are considered, then a broad sum of money can and should be allocated within the maintenance budget to cover such eventualities.

## 10.7 The Importance of a Programme

Having a programme in place is of paramount importance when managing the maintenance of buildings. A programme is essentially a document that visually outlines the full scope of project requirements on a sequential basis and in a chronological order. Time will usually be portrayed along the X-axis and the list of activities along the Y-axis. Each required activity is then plotted on the programme against the amount of time allocated for its completion. If the starting of one activity is dependent upon the completion of another, then these activities can be linked, and if there is a sequence of linked activities that the overall duration of the programme is dependent upon then these are called the critical path. A delay to the critical path will ultimately mean a delay to the overall project duration.

Often a programme is treated as a recording exercise to reflect what has happened. Whilst this exercise does have benefits, it is important to view the programme as an ongoing and 'live' document that is constantly reviewed and updated. This should also not be a solo task. Whilst it will come down to one professional to update and review the programme, its success ultimately comes down to a collaborative multi-party approach. The specialist skills sets and knowledge of a wide range of professionals and operatives will be required to ensure a programme's relevance and accuracy.

All maintenance activities should be plotted on a programme and then this will serve as a checklist for both the tasks required to be undertaken and the expected duration of each activity. The full building maintenance team will then be able to work from the same document and all be aware of what tasks need completing and when. This can help the focus and management of resources and to ensure all tasks are completed in the correct and most pressing order. There are numerous benefits to having a maintenance programme in place, including:

- Progress against the programme can be mapped to ensure delays are identified as early as possible so they can be minimised and contingency plans put in place.
- Work start and end dates, as well as expected durations, can be forecast so that expectations of impacts and any disturbances can be planned for and managed.
- If any activities require re-sequencing then having a programme in place can allow for this to happen effectively. For example if a planned proactive maintenance event has turned into a reactive immediate event due to a deterioration in equipment quality then the activity can be moved forward on the programme and the impact this will have on the scheduling of other activities can be easily identified.
- A culture of collaboration can be promoted across different disciplines as everyone provided input, and requirements can be acknowledged and represented in a programme.
- Resource planning can also be carried out effectively, with activities moved to different times so that the current levels of employees remains optimum to address the required maintenance issues. Similarly, times when current employee levels will not be enough to meet maintenance demands can also be easily identified and appropriate recruitment action undertaken.
- Operative training needs can also be identified in examples where servicing requirements are usually handled in house and a bespoke piece of equipment requires maintenance. Existing operatives can be identified to complete the required servicing and so an effective schedule of when the servicing is required will allow the operatives to be professionally trained to conduct any servicing needs.

In order to develop an effective programme, or to input into the development of a programme as part of a collaborative team, an individual will not need to have an extensive range of skills, knowledge, and experience, but simply a willingness to collaborative and the ability to think clearly with regards to the requirements of each maintenance activity. However, the more skills, knowledge, and experience an individual has, the easier they may find it to contribute to a programme's development. In order for a comprehensive and robust maintenance programme to be developed, the following are crucial elements:

- An understanding of the importance of dates so that bank holidays and international events that may cause supply chain equipment delays are fully understood.
- An appreciation of task requirements so accurate durations can be included, and the programme can serve as an accurate reflection of the works required to be undertaken.

- An awareness of key stakeholders and their specific requirements of the project and the programme, and of how the programme can be best used to assist each stakeholder group in achieving their aims and the aims of the building.
- A full knowledge of contract operation and administration principles so that all contract requirements can be incorporated into the programme and the programme can serve to accomplish and contribute to many contract requirements.
- An awareness of the building, and its unique needs and requirements, including any events that will be held and uses the building will need to perform to ensure all works are scheduled for times that suit the building and occupant requirements.
- Regular meetings will also be required to ensure the programme remains relevant and is updated at regular intervals to reflect any progress made or changes to the building's requirements.
- An understanding of the building and maintenance task risk events that may influence how the building is used and how the planned works are completed.

## 10.8   Site and Task Constraints

Once a programme has been developed and shared, and a project commenced, site constraints may be encountered when it becomes time to undertake the required maintenance work. Site constraints should always be considered during the development of the programme, as they may impact on the anticipated duration of any tasks. However, sometimes the full nature of any site constraints does not become visible until contracts are signed and a project commences. This is why, for any new building maintenance projects, it is essential to build inspection opportunities into the programme of schedule works, well in advance of the due maintenance dates. Firstly, such inspections will reveal the urgency of any maintenance works that need to be undertaken. Secondly, accurate task requirements can be assessed, and any site or task-specific constraints can be identified.

This is an important step, and the role inspections play cannot be overestimated. As inspections are often time-consuming and can be costly, they are sometimes not utilised, but this is going to leave the building maintenance professionals exposed to increased risks of additional delay and expense being encountered when the full maintenance works are being undertaken. For example, a gas contractor who is responsible for the maintenance of several gas assets (such as governor units and overground pipework) may be engaged in a procurement process whereby the client wants a fixed price and time frame for the maintenance works to be completed. It would therefore not be wise for the contractor to price up the works and the time frame involved based purely on the information provided by the client at tender stage. A full site inspection of each applicable asset would be the only sensible course of action so site constraints for each asset can be fully understood. This will allow full time frame requirements to be known and a more accurate cost to be submitted covering all site requirements. Imagine if one asset was surrounded by overgrown vegetation, or if another was only accessible via a field or narrow footpath. The first may require operatives to spend additional time clearing the site before works can commence and the second may limit the use of any usual plant and equipment, resulting in specialist equipment needing to be hired in or requiring increased levels of manual handling. All are examples of site-specific constraints that can add to the time and cost involved in maintenance works.

Site and task constraints can also be more indirect, such as ability to access a certain piece of land at certain times. For example, for any significant building maintenance works at a school, a school car park may be off limits for the storage of materials, situating of skips, and parking of work vehicles during term time, but over the Easter or summer holidays such uses would be permitted. Window cleaning and cladding repair may be another maintenance issue where a site inspection would be beneficial. Three elevations of a building may be easily accessible by a tarmac road and car park, but the fourth elevation that is only visible from the rear of the building may reveal that there is a sloping grass bank and only a narrow stone path for access. Therefore, the plant and equipment that will be optimum for undertaking the required maintenance work on three elevations will not be suitable for the fourth. Being unaware of such factors when starting maintenance works will require reactive solutions that may negatively impact the project time and cost. Inspections may cause initial expense and delay in starting any 'actual' maintenance works, but they are invaluable in assessing and understanding site and task constraints. They are therefore essential to ensuring maintenance works are completed in not only the most timely and cost-efficient manner, but an awareness of the full requirements helps to ensure compliance with the highest health and safety standards.

## 10.9   Health and Safety of Building Maintenance

Health and safety (H&S) is an encompassing term that refers to the behaviours, procedures, strategies, obligations, and legal requirements aimed at preventing injuries. Everyone is H&S conscious on some level in all situations and environments. When crossing the road for example, most will stop and check for oncoming traffic. When a light bulb needs changing in a high-ceiling room, rarely will anybody stand three chairs one on top of the other and then proceed to climb up them all to reach the light. We have an ingrained H&S awareness, to look out for, and avoid, danger. When considering the H&S of building maintenance, it is simply the transferring of this focus on staying safe to the requirements of building maintenance, and ensuring the requirements can be met, whilst safety maintained.

Building maintenance largely occurs by employees of a company operating for a client and classed as a workplace, and therefore legal requirements such as the Health and Safety at Work Act (1974), sometimes referred to as HASAWA, govern behaviours of those involved. HASAWA is the main piece of legislation concerning occupational H&S and covers the duties employers have towards their employees, the duties they have towards members of the public, and the duties employees have to themselves and others. H&S during maintenance works is of the upmost importance, as accidents occurring during the undertaking of maintenance works significantly contribute towards annual accident statistics in the UK. However, there are several broad initiatives and processes that can be adopted as part of building maintenance H&S in an attempt to minimise accident prevalence and reduce accident severity.

### 10.9.1   Having Thorough Supply Chain Selection Methods in Place

It is reasonable to utilise suitable pre-qualification questionnaires with all potential supply chain partners to ensure their basic competence and understanding of the risks involved. This should then be built upon further with a set of minimum H&S standards and requirements that all supply chain partners must adopt. Therefore, you can have faith that any supply chain partner you approach to submit a cost meets your company's H&S requirements and so is fit and proper to carry out the work if awarded.

### 10.9.2   Operating a Permit to Work System

To ensure that maintenance work cannot begin at an incorrect time, or without all stakeholders being aware, a permit to work system can be in operation. This requires those undertaking the maintenance work to apply for and be provided a permit. This permit will stipulate who can carry out the work, where the works are to be completed, and what dates the permit is valid for. This will then prevent others from being in the area at the same time, as once a permit is issued, information regarding that permit can be communicated to all appropriate stakeholders.

### 10.9.3   Ensuring Inspections Are Carried Out for All Required Works

As covered in this chapter, inspections are essential to allow the full requirements and ramifications of a task to be understood. They also allow for thorough planning and preparation to take place so that operatives are fully informed of the site conditions and will not be surprised by what they encounter once the task begins. Having knowledge of the site conditions will allow for hazards and sources of potential danger to be identified and suitable risk assessments and method statements be drawn up to enable the work to be completed in a safe manner.

### 10.9.4   Ensuring the Risks Are Fully Understood

The risks that accompany activities also need to be understood by each operative undertaking them. It is unfair, unsafe, and ethically incorrect (as well as illegal) for an individual to engage in an activity without fully understanding the risks involved. If a risk is unknown at the outset, despite the best practice being followed, then that is different. It is still a situation that needs to be dealt with; however, it is not the same as starting an activity unaware of the risks because adequate inspection and planning has not been undertaken. Only once the risks are fully understood, including any mitigation practices in place, and the risks that remain, can an operative begin to undertake the maintenance works required.

### 10.9.5   Being Up to Date with the Latest Guidance and Legislation

Ensuring that the latest legal standards and industry best practice guidance documents are understood and adhered to is of paramount importance, as materials that were once considered safe may have been recalled due to prevalent issues of malfunction. And so such materials must not be used on any future maintenance works, even if a large supply of them still remains in a building's stock. When maintenance work is conducted, it will also need to be in line with all latest standards and legislative requirements. Having no knowledge of such requirements is not a reasonable justification as to why the requirements were not adopted, and in such circumstances can leave an individual building maintenance professional, as well as their employer, open to liability.

### 10.9.6   Having an Up-to-Date Training Matrix

A thorough record needs to be kept of all operatives' current training, along with all durations the training is valid for, and all expiry dates. This is so qualifications and training can be gained before the expiry of early course completions, and also so that there is a clear record of which operatives can conduct which maintenance elements. If gaps are identified and no operative is suitably trained to conduct a certain element of maintenance work that will be required, there will be enough notice to get the relevant people trained before the scheduled maintenance. Failure to do so, and an insistence that operatives complete tasks they are not trained to do, will result in people operating beyond their competence levels, which would then increase the likelihood of accidents occurring.

## 10.10   Common Difficulties Encountered during Maintenance Works

There are many common building maintenance works, and therefore many common difficulties encountered when trying to assess and address such required works. Chapter 3 includes some of the most common maintenance issues experienced by building maintenance professionals, but there are also common issues that negatively impact such professionals when trying to assess the conditions of existing building to calculate maintenance requirements. This can include issues with neighbours and neighbouring properties whose land (or air) you may need to utilise for temporary plant and equipment or for access or legal covenants and easements that apply to the building you are maintaining, local restrictions on noise and transport, or even the proximity of the building you are maintaining to local nature reserves and Areas of Outstanding Natural Beauty (AONB). One major area of difficulty that is more prevalent with older buildings and assets is the lack of clarity over any previous work that has been carried out, and the lack of detail over original construction and engineering processes used. Additional time, costs, and resources are often spent ascertaining the construction materials used in a building, or surveying the strength of beams and columns, when ideally this information could be made available by the main contactor or client if it had been recorded and stored correctly. For newer buildings and assets such good practices are now widely adopted to prevent future unnecessary surveys and inspections being carried out, and a process of 'soft landings' is common practice.

## 10.11   Soft Landings

Traditionally, a building would have been constructed, and then once complete it would have been handed over to the client. The client would then collect the keys and have access to their building – but they may not be familiar with all the building's features. The operation of all the equipment, or the correct maintenance procedures to follow to maximise the equipment's performance, or why certain decisions had been made with regards to materials and equipment used. This would then ultimately result in the building not being used as intended or maintained as required and so its performance would not meet expectations. The contractor could offer to show the client all the features they require, and this would help towards ensuring the building is used correctly and so its performance is maximised. However, dependent upon the building size and purpose, this process could take anything from a day to a few weeks. If this occurs once the building has been handed over, then the client will not be able to fully use the building from day one and will not be able to maximise potential returns. If this process of training and educating the client were to take place during the construction phase instead, it would mean the client could maximise their building as soon as possible after handover; however, it could lead to delays in completion of the building, as the main contractor is focusing resources on client education over building completion.

This is where 'soft landings' are utilised. As opposed to the 'hard landing' that may be felt by a client who takes their keys post completion and then has to figure out how their building works, a soft landing is a structured transition strategy agreed between the main contractor and client. It will be agreed at the tender stage of the project and involve a schedule of training and test days and events that start occurring at completion nears, so that when completion is achieved the client is fully aware of how to ensure optimum building performance. It is not always feasible that building maintenance professionals are involved at the tender stage of the building construction, and so involvement at an early stage rarely occurs. Indeed, building maintenance professionals are often not engaged until the client has possession of the building and so do not even become involved during the construction stage. This is a great shame, and something that initiatives within the industry are hoping to address. It does not occur in all instances, and where building maintenance professionals are engaged by the client during the construction phase, they are in attendance at the soft-landing events and so are able to assist the client in maximising building performance.

There are numerous frameworks and strategies that are available that can assist in the structuring of soft landings, and to help ensure they are executed effectively so maximum benefit is realised for all those involved. Arguably, the ultimate aim and intention of them all is the same, to help the client and building end user maximise the building performance and be fully confident in having knowledge of the building operations and requirements at the earliest possible stage to have a positive user experience.

The requirements for soft landings are growing, and in 2016 all centrally procured UK government projects needed to abide by the Government Soft Landings (GSL) strategy. The intention of this strategy was for the construction stage to retain a focus on the client's requirements and the end user's requirements for the building, which can often become forgotten in the race to complete the project on time and within the agreed budget. In addition to the soft-landing initiatives in existence, those with responsibility for completing construction projects also compile and handover operation and maintenance manuals upon completion of the building.

## 10.12 Operation and Maintenance Manuals

Operation and maintenance manuals – or they could be referred to as health and safety files, or building owner's manuals – are the instructions and operating procedures combined in a single location for the benefit of a building's owners, managers, and users. They should be created and compiled during the construction phase of a building, with usually the main contractor taking responsibility for them. There could be either a hard copy or electronic copy, or often both.

1. The first step is to create a database of all materials, subcontractors and consultants involved in the building's construction.
2. Either a third-party subcontractor can be procured or the creation of the manuals can be completed 'in house'.
3. Within the database created in step 1, a list of all materials and permanent equipment in the building should be made, as well as the professional services provided.
4. Information on each of these items should then be requested from the appropriate party and filed in an easily accessible and logically structured manner.
5. Once complete, the required amount of copies of the manuals should be made and distributed to the client with the handover of the building.

These manuals will then be retained by the owner, and a copy should be provided to the building maintenance professionals. This copy can then be referenced during any repair work. This is of paramount importance if, for example, a building material that is currently widely used today but in future is classed as hazardous; the extent of its use can be easily understood when any repair works are carried out so that future operatives can be safeguarded as effectively as possible. If a material requires to be replaced, such as a carpet tile or suspended ceiling tile, then the manuals can also be an effective first point of call. The brand, model, manufacturer and installer details should all be available so the exact product can be sourced as a replacement for the defective one, and the same company could even be contacted to install if the defects liability period has passed.

Each consultant and engineer involved in the design and development of the building will also have an input in the creation and compilation of the files, as all drawings and calculations will need to be included in case future extension work or additional building works are to be undertaken. The original building details will be required to ensure additional work can be carried out in an effective and efficient manner, with minimal unnecessary disturbance. Being aware of the construction of all building elements can help this occur, as future architects and structural engineers will be able to assess the current building with increased confidence, as they have knowledge about how and why it was originally constructed as it was.

The format of the operation and maintenance manuals is currently going through a period of change. Whilst the fundamental principles remain intact (that of compiling and capturing knowledge at the construction stage for the benefit of the operation stage), the process by which this takes place can now be part of the overall industry move towards building information modelling (BIM).

## 10.13   Building Information Modelling

For many people, building information modelling (or management) (BIM) represents a 3D model that contains all relevant information regarding a building. This is not the case. Whilst this short section does not intend to act as a holistic guide to BIM (plenty of textbooks and guides currently exist that focus purely on BIM), this section does serve to offer a brief introduction to BIM, and how it can inform the decisions of building maintenance professionals.

BIM ultimately refers to how stakeholders manage digital information. Its aims and intentions are to encourage collaboration, enhance communication and ensure that shared creation of information in a digital format takes place. This collaboration is not limited to purely the construction phase of a building, but should start at the inception stage, and continue through the construction, operation building maintenance, refurbishment, and demolition. This is where building maintenance professionals often have their hands tied as many of the decisions regarding the adoption of BIM principles are taken prior to their involvement. Nevertheless, if BIM principles are adopted, it can lead to a shared centralised digital vault of building information (digital twinning of the building) that can be accessed and updated by all appropriate stakeholders throughout the building's life.

You can view BIM as essentially a table. This table is actually an online shared platform known as a content management system. Each stakeholder (client, architect, consultants, main contractor, subcontractors, building maintenance professionals, end users, etc.) then has a seat around this table (access to the shared online platform). Instead of the traditional mentality of each stakeholder operating in a 'silo' when it comes to data creation (i.e. the architect creating a drawing and then sending it onto the engineer, who then adds the calculations before sending it on to the main contractor/quantity surveyor who will then send it to the main contractor), BIM encourages all the stakeholders 'sat at the table' to create this data centrally and to 'share' the same live data and project information. All specifications will be drafted together 'on the table' if you will, and the building information requirements, including drawings and specifications, will then be created together, so that any clashes can be detected immediately, and issues proactively resolved, well before work starts on-site. For building maintenance professionals, this level of knowledge and input can be invaluable.

If, at the touch of a button, a building maintenance professional can access the entire history of a building via a centralised and collated database (the digital building twin) where each piece of information was collaboratively produced, then the accuracy of this information can be counted upon. It will also save time that is all too often spent trying to locate previous records or having building elements surveyed so structural aspects and previous construction methods can be ascertained. This information (and much more) will be available in whatever software platform is utilised for BIM. The concern, however, is that the adoption of BIM as an approach to collaborative data could not include input from building maintenance professionals at an early enough stage, and so would not contain the relevant and required information. The proposals for BIM (if adopted fully) do seek to appease these concerns and will make the role of the building maintenance professional easier and more time efficient.

The 3D model that many people associate with BIM can be the 'table' that is described above. BIM, however, is not dependent upon the use of a 3D model to work, and it is the actual open plan sharing and creative collaboration of a project's data to help inform decision making during the construction, operation and demolition stage of a building. The principles of BIM can be adopted to different levels, which have been categorised as 'maturity levels' and are:

- Level 0 – Minimal collaboration whereby only 2D drawings are produced. These drawings and text documents are communicated either in print or electronically.
- Level 1 – Little collaboration, but a mixture of 2D and 3D information is created and shared electronically via a common data environment (CDE).
- Level 2 – Collaborative working takes place. Electronic 3D data produced that exports a common file format. Data is still created in silos, but 'shared' models can be created from the data produced. Programme and cost information could also be included in the drawing details. This is the standard required for UK public sector projects.
- Level 3 – Fully collaborative project data including programme, cost and life cycle information. Models and data are created together and fully shared amongst the project team.

## 10.14 Conclusion

Managing the maintenance process is a difficult undertaking. The variety of tasks that are required to be competed and numerous processes to be followed, frameworks to be adopted, standards to be achieved, legislation to abide by, time frames to complete within, budgets that cannot be exceeded, and quality checks that must be adhered to, are an ever-changing ocean that must be navigated. This navigation falls to the building maintenance professional. This chapter serves to give a brief overview of just some of the requirements of the role, and how the requirements of the role can be effectively managed. Many other requirements of the building maintenance professional can be found within all the other chapters of this book. This chapter attempted to focus on a few key broad lessons that building maintenance professionals can adopt to help ensure building requirements are achieved in an effective and efficient manner.

The importance of proactive maintenance schedules cannot be underestimated, nor can the benefits that are experienced when comprehensive programmes are created in a collaborative and supportive multi-stakeholder environment. The advantages that early inspections provide can also not be discounted, nor can the efficiencies achieved across all metrics if inspections are adopted. How site constraints can be managed and the simple, yet effective, role health and safety principles can play in ensuring buildings are maintained at optimum levels is also discussed. Having thorough supply chain practices, operating permit to work schemes, ensuring regular inspections are undertaken, ensuring that risks are fully understood, that there is an awareness of the latest guidance, and being aware of the training and competence of your own workforce are all simple management practices to have in place but can yield untold benefits to the management of maintenance projects. Finally, the idea and principles behind soft landings is discussed, due to its current and growing importance within the industry, as well as the transformative role BIM can play for the building maintenance professional, and the benefits that all stakeholders can achieve if the building maintenance professional is engaged with at an earlier stage in the planning and construction of a building.

# 11

# Conclusion

The increasing use of private finance for the delivery of major public infrastructure together with the publications of various policy pronouncements as part of the UK government's Industry Strategy for Construction 2025 has caused a paradigm shift in the construction industry's view of the maintenance and upgrading of existing buildings, with asset management no longer seen as the 'Cinderella' of the construction industry. As a result, higher education providers have started to place an increased focus on the longer-term view of construction projects advocated in Wolsteholme's critical review of the construction industry, published in 2009 on behalf of Constructing Excellence, in which he attested that construction projects cannot be seen in isolation from the operational phase of the completed building.

Indeed, ongoing movements within the built environment are aimed at addressing this long-standing disconnect between how buildings are constructed and how they are then used and maintained. To an industry outsider, the manner in which the construction and then the maintenance of buildings are viewed as two entirely separate entities must appear strange. This book attempts to support the bridging of these functions (building construction and building maintenance) and illustrates how the early engagement of a building maintenance professional can serve to enhance the construction process as well as benefit the client when the building is completed, handed over and fully operational.

This book started by outlining the importance of being able to survey existing buildings. It is key to understand the requirements of surveying buildings correctly, as it is only through the timely and accurate diagnosis of defects can the full ramifications of required maintenance work be understood. Chapter 2 addresses the issue of misdiagnosis but with practical examples and scenarios outlined, the reader will understand the skill sets and knowledge required to ultimately identify and treat signs of building deterioration. Effective communication is a common theme throughout this book and one of the main lessons all building maintenance professionals should learn. The lessons on effective communication start in this chapter, with the benefits of clear and early client engagement outlined. The importance of understanding the requirements of clients and ensuring the correct type and nature of information is suitably captured is explained. The capturing of this information can take many forms, and this chapter outlines the different types of building surveys available, a must for all those new to the role of building maintenance professional.

*Introduction to Built Asset Management*, First Edition. Dr Anthony Higham, Dr Jason Challender, and Dr Greg Watts.
© 2022 John Wiley & Sons Ltd. Published 2022 by John Wiley & Sons Ltd.

Chapter 3 provides an insightful overview of the common maintenance issues and building upon the knowledge of the earlier chapter, takes the reader through the ways and means of identifying and addressing building defects. The key lessons of understanding true causes and ramifications of defects and having the ability to analyse underlying conditions is discussed in this chapter. It is imperative the most suitable form of repair is understood and then carried out in a time-efficient manner and this chapter provides the skills required to ensure this can be achieved. Planned and reactive maintenance is then first introduced in this chapter and the importance of early building maintenance professional engagement discussed. This is an ever-important issue that appears throughout this book in an attempt to support the small ripples in industry that are happening with regards to understanding the importance of capturing building maintenance knowledge at the development and construction stages of a building's life cycle.

Maintenance management and performance measurement as part of PFI (private financing initiative) schemes is then covered in Chapter 4. At the heart of contemporary building maintenance, indeed, at the heart of modern-day business, is performance management. This chapter introduces the importance of performance management as a focus point in current building maintenance requirements and expectations and how an awareness of performance management is directly linked to service and quality improvements. For those involved in facilities management, this chapter also covers the main issues facing PFI schemes and outlines the more nuanced issues facing those working in the healthcare sector who have been tasked with securing more for less when it comes it operational expenditure. The key lessons of this chapter are a focus on performance measurement with clear accountability, and a proactive approach can help achieve increased quality and ensure cost-efficiency savings are achieved. The skills and practices the building maintenance professional needs to achieve these outputs is clearly discussed as is the role payment mechanisms can play in ensuring the desired quality levels are achieved.

The importance of procurement and contract administration in building maintenance is covered in Chapter 5. The key principles are clearly articulated for the benefit of building maintenance professionals and how these principles can be applied to refurbishment works are outlined. By having an understanding of the appropriate procurement routes, professionals can make informed and correct decisions based on the information they have that best serve the needs and requirements of the building and its stakeholders. Strategic procurement knowledge is an essential tool in the arsenal of the building maintenance professional and this chapter covers the main aspects to inform and accompany decision making. Making the correct procurement decisions is, however, only half the battle. Deciding upon the correct contract and then administering it as intended is how project success is truly achieved. Such topics form the second half of this chapter. Drawing upon industry surveys, reports and market analysis tools, this chapter concludes by discussing the main contract options and their appropriateness to different maintenance situations. Ultimately, the lesson of this chapter is that making correct and informed decisions when it comes to selecting suitable procurement routes and contract options will help the management of maintenance projects. Incorrect decisions will lead to an increased level of disputes and a lower likelihood of all project objectives being achieved successfully.

The important, yet often overlooked, ideas of managing operation expenditure in the refurbishment and maintenance marketplace are covered in Chapter 6. Such areas contribute a significant amount to the value of the annual construction expenditure and is set to increase over the coming years. The requirements for maintenance and refurbishment work are vast, and some of the most popular client motivations are discussed so that awareness is increased and common maintenance issues can be addressed in a more formal and thorough manner, building upon previous lessons learnt to ensure projects are completed more efficiently. The key lesson of this chapter is that pre-contract financial management is a precursor to effective project financial management. The purpose of cost plans and how they are crafted and developed are covered, as this is an essential skill for building maintenance professionals who are involved in commercial and facilities management.

Chapter 7 provides an overview of life cycle cost analysis. This is a concept growing in importance for both clients and contractors, and building maintenance professionals need to be at the forefront of this paradigm shift as the industry moves away from a lowest priced model to focus on value achieved over the medium and long term. The trade-off between capital costs and operational costs is now a real consideration for clients, and the building maintenance professional needs to be on hand to provide up-to-date and comprehensive advice to the industry. Budgeting for maintenance works is also introduced in this chapter and the concept of TOTEX – total expenditure (capital plus operational) – is discussed, and how this could provide the answer to those who focus on simply on OPEX (operating expenses) and those who focus solely on CAPEX (capital expenditure).

The key lessons to take from Chapter 8 are that the expectations for a more sustainable construction industry, and sustainable building maintenance processes, are upon us. All stakeholders now expect as standard practice the most sustainable course of action to be taken, and the principles of the circular economy and corporate social responsibility to be embedded within companies. Therefore, these concepts are introduced and discussed and how they relate to building maintenance projects. The carbon neutrality movement, along with the ideas behind, and benefits of, retrofitting are also outlined and how and why building maintenance professionals should not only be aware of these concepts but be actively looking to embrace the principles at every opportunity. The ambiguity that surrounds some ideas is then discussed, and how the introduction of the Sustainable Development Goals (SDGs) has sought to ensure all individuals and organisations are pulling in the same direction and trying to achieve the same goals. The contribution building maintenance professionals can make to achieving the SDGs cannot be underestimated, yet this is an area that is not explored in much detail in the literature or in current practice. It is hoped this will change in future and the positive impact building maintenance professionals can have across many of the SDGs is recognised, not least by many building maintenance professionals themselves.

The concept of risk management is introduced in Chapter 9. This chapter starts by introducing the idea and nature of risk and how risk is generally viewed within the built environment. This chapter aims to explain how the management of risk can be successfully undertaken within building maintenance and asset management. The common events that can give rise to such risks are also explored and how the consequences of these events can be understood and hopefully their likelihood and severity both reduced. Correct and suitable risk management should be standard practice, but it is all too often not. This chapter discusses the opportunities that risks present and the reasons why some professionals may choose to not engage with risk management practices effectively. Ultimately, this chapter provides the information behind why both proactive and reactive risk management is required, and the benefits that a building and the building maintenance team can experience if the time is taken to understand the risks involved and plan for their effective management.

Finally, Chapter 10 discusses the management of the maintenance process. The chapter covers the broad topics of proactive and reactive maintenance, and the importance of each. The key role programming and planning plays in ensuring maintenance works are undertaken in an efficient and effective manner are outlined, as it is of paramount importance building maintenance professionals are aware of the multiple benefits that can be easily achieved with an effective programme in place, that is viewed as a live and ongoing document, and updated on a regular basis by a collaborative and multidisciplinary team. The task and project constraints operatives face when carrying out maintenance works are also discussed and these lead into a discussion of the broad health and safety management aspects applicable to building maintenance. Concepts such as soft landings are also introduced. The smooth transition between the construction and operation phase of a building needs to be encouraged and to become more prevalent in the life cycle of buildings. Only then will the performance of buildings start to increase and meet the expectations clients have based on the projections during design and construction. This chapter ends with a brief and basic overview of building information modelling and the benefits this can bring, especially with early engagement of the building maintenance professional. If you work in construction, you need to be engaging with building maintenance professionals during the design and construction stages. If you are a building maintenance professional, the earlier you can become involved in the design and construction stages the more successful the project will be, the more the clients aims will be achieved, and the greater the performance of the building during operation.

# About the Authors

**Dr Anthony Higham BSc, PGCE, MSc, PhD, MRICS, MCIOB, FHEA** is the Head of Construction and Management at the University of Salford. Anthony is an experienced Quantity Surveyor and Project Manager having worked in the construction industry for 10 years before moving into academia 15 years ago. Anthony has published extensively in the field of Construction Management authoring two previous textbooks, book chapters and over forty journal and conference papers focused on construction economics, procurement and social sustainability with a focus on the measurement and evaluation of sustainable benefit delivery at project and organisational levels.

**Dr Jason Challender MSc FRICS FAPM FAHE** has acquired over thirty years 'client side' experience in the UK construction industry and procured numerous successful major construction programmes during this time. He is Director of Estates and Facilities at the University of Salford, member of its Senior Leadership Team and responsible for overseeing a large department of approximately 350 estates and construction related staff. One of his major recent achievements includes raising the sustainability credentials of the University of Salford to becoming Eco Campus Platinum and achieving the highest accreditations for carbon management and energy efficiency in the last two years. He is also a construction academic researcher with four books and eleven academic journal and conference papers published in the last few years, all of which have been dedicated to his studies around the construction industry. Furthermore, he has previously participated as a book reviewer for Wiley. He has also attended many national construction and institutional conferences as a guest speaker over the years. He is a Fellow of three institutions including the Royal Institution of Chartered Surveyors (RICS), Advance Higher Education (AHE) and the Association of Project Managers (APM). Furthermore, he is a Board Director of the Royal Institution of Chartered Surveyors and the North West Construction Hub.

*Introduction to Built Asset Management*, First Edition. Dr Anthony Higham, Dr Jason Challender, and Dr Greg Watts.
© 2022 John Wiley & Sons Ltd. Published 2022 by John Wiley & Sons Ltd.

**Greg Watts BSc, PGCert, PGCAP, EngD, MRICS, FHEA** is the Director of Quantity Surveying at the University of Salford. He is also an experienced Quantity Surveyor having worked extensively in the construction industry for over 15 years prior to moving into academia. He has previously published two textbooks, and numerous book chapters, conference papers, and peer reviewed journal papers, spanning the subjects of procurement, sustainability, social value, Corporate Social Responsibility, infrastructure, project success, education, and Continuous Professional Development.

# Index

*Introduction to Built Asset Management*, First Edition. Dr Anthony Higham, Dr Jason Challender, and Dr Greg Watts.
© 2022 John Wiley & Sons Ltd. Published 2022 by John Wiley & Sons Ltd.